Nuclear Power and the Spread of Nuclear Weapons

ALSO BY THE NUCLEAR CONTROL INSTITUTE

Nuclear Terrorism: Defining the Threat
Edited by Paul Leventhal and Yonah Alexander

Preventing Nuclear Terrorism: The Report and Papers of the International Task Force on Prevention of Nuclear Terrorism
Edited by Paul Leventhal and Yonah Alexander

Averting a Latin American Nuclear Arms Race
Edited by Paul Leventhal and Sharon Tanzer

The Plutonium Business
By Walter C. Patterson

The Tritium Factor
Proceedings of a workshop co-sponsored with the American Academy of Arts and Sciences

Nuclear Power and the Spread of Nuclear Weapons

CAN WE HAVE ONE WITHOUT THE OTHER?

EDITED BY PAUL L. LEVENTHAL,
SHARON TANZER,
AND STEVEN DOLLEY

FOREWORD BY
REP. EDWARD J. MARKEY (D-MASS.)

BRASSEY'S
Washington, D.C.

Library of Congress Cataloging-in-Publication Data

Nuclear power and the spread of nuclear weapons : can we have one without the other? / edited by Paul Leventhal, Sharon Tanzer, and Steven Dolley.— 1st ed.

p. cm.

Includes bibliographical references and index.

ISBN 1-57488-494-8 (cloth : alk. paper) — ISBN 1-57488-495-6 (pbk. : alk. paper)

1. Nuclear engineering—International cooperation. 2. Technology transfer. 3. Nuclear weapons. 4. Nuclear nonproliferation. I. Leventhal, Paul. II. Tanzer, Sharon, 1941– III. Dolley, Steven.

TK9145.N8297 2002
327.1'747—dc21 2002018416

Printed in the United States of America on acid-free paper that meets the American National Standards Institute Z39-48 Standard.

Brassey's, Inc.
22841 Quicksilver Drive
Dulles, Virginia 20166

First Edition

10 9 8 7 6 5 4 3 2 1

It is not certain that the public will accept an energy source that produces vast amounts of radioactivity as well as fissile material that might be used by terrorists. *Enrico Fermi*

There is no prospect of security against atomic warfare in a system of international agreements to outlaw such weapons controlled only by a system which relies on inspections and similar police-like methods. The reasons supporting this conclusion are not merely technical, but primarily the insuperable political, social, and organizational problems involved in enforcing agreements between nations each free to develop atomic energy but only pledged not to use bombs.

Acheson-Lilienthal Report, 1946

Nuclear is different. *Theodore Taylor*

CONTENTS

ILLUSTRATIONS

TABLES

FOREWORD

On the twentieth anniversary of the Nuclear Control Institute, it is a singular honor to have been invited to write the foreword for this timely and important book. It brings to the public at large the excellent presentations and discussions that took place on April 9, 2001, at the twentieth anniversary conference sponsored by the institute. The title of the conference, "Nuclear Power and the Spread of Nuclear Weapons: Can We Have One without the Other?," posed a very important question: Does the proliferation of nuclear power encourage the proliferation of nuclear weapons? The conference provided an exceptional opportunity to review the connections between nuclear power and nuclear weapons and to examine critically the desirability of increased reliance on nuclear power when the risks of proliferation might outweigh the energy benefits. The conference was a testament to the Nuclear Control Institute's persistent dedication to the cause of keeping the world safe from the potential dangers of nuclear materials.

The Nuclear Control Institute was the first nonprofit organized to work exclusively on the problem of nuclear proliferation. The institute's focus has always been the prevention, and not simply the management, of the further spread of nuclear weapons. It works to eliminate civilian uses of atom bomb materials, plutonium, and highly enriched uranium by calling attention to the dangers they pose. The institute seeks to break the linkages between civilian and military nuclear applications and to build linkages between nuclear disarmament and nuclear nonproliferation. However, the Nuclear Control Institute has always made clear that it is not a priori averse to nuclear power. It is averse to nuclear power's being used improperly and irresponsibly, and it wants the United States to be the leader in its responsible use.

For twenty years Paul Leventhal and the Nuclear Control Institute have been working to safeguard this country from the dangers of irresponsible and malicious use of nuclear materials. For an organization of modest size, its accomplishments have been great. I want to highlight several of them.

1. In 1982 the Nuclear Control Institute fought successfully to enact the Hart-Simpson-Mitchell amendment, which banned the use of spent fuel from U.S. civilian nuclear power plants as a source of plutonium for weapons.
2. In 1985 the institute fought for tough nonproliferation conditions to be attached to a resolution approving the nuclear cooperation between the United States and the People's Republic of China.
3. The institute's 1985 international conference on the threat of nuclear terrorism led to the establishment of the International Task Force on Prevention of Nuclear Terrorism. The task force's findings in 1986 contributed to enactment of the Omnibus Diplomatic Security and Anti-Terrorism Act of 1986.
4. In 1987 the institute helped win enactment of the Murkowski amendment, which blocked air shipments of plutonium from Europe to Japan, after the institute disclosed the failure of a test to prove the crashworthiness of a plutonium shipping cask, a result that had been kept secret from the public.
5. In 1989 the institute convened a conference of Argentine, Brazil, and U.S. nuclear officials and experts that developed proposals that Argentina and Brazil incorporated into the treaty they signed the following year to end their nuclear arms race.
6. In 1989 the institute fought for passage of the Markey amendment to the Tiananmen Square sanctions legislation to close loopholes in nuclear export control laws relating to China.
7. In 1990 the institute correctly predicted that Iraq would violate the safeguards of the International Atomic Energy Agency and divert civilian nuclear research reactor fuel for the production of nuclear weapons.
8. In 1991 the institute fought successfully for House passage of the Markey-Wolpe amendment to the Export Administration reauthorization to require that exports of dual-use items on

the nuclear referral list be contingent on a nuclear cooperation agreement between the United States and the importing country.

9. In 1992 the institute helped gain passage of the Schumer amendment, which barred U.S. transfers of highly enriched bomb-grade uranium to research reactors that could make use of newly developed, low-enriched uranium fuel unsuitable for weapons. U.S. exports of highly enriched uranium are now virtually eliminated, and most reactors in Europe not using low-enriched fuel have agreed to convert to it.
10. After a decade of effort, the institute and the Committee to Bridge the Gap finally persuaded the Nuclear Regulatory Commission to promulgate in 1994 a rule to protect reactors from truck bombs.
11. Since 1996, the institute has played a leading role in advocating safe and secure disposal of military plutonium directly as waste and opposing its use as a civilian reactor fuel in the United States and Russia.
12. In 2001 the institute supported the Markey amendment to require the Nuclear Regulatory Commission to upgrade the design basis threat, codify the Operational Safeguards Response Evaluation program, and mandate presidential assessment of the appropriate federal government role in the defense of nuclear power plants against terrorist attack.

Leventhal's involvement in these critical issues actually predates the emergence of the Nuclear Control Institute. For years before he formed the institute, he played an absolutely crucial role as a Senate staff member in helping to replace the Atomic Energy Commission with separate regulatory and promotional nuclear agencies, to enact the Nuclear Non-Proliferation Act, and to direct the Senate investigation of the Three Mile Island accident—three critical milestones in the management of nuclear affairs in the 1970s and early 1980s.

Since then, Leventhal and the Nuclear Control Institute have fought the good fight against development of a full plutonium fuel cycle, particularly to end civilian commerce in, and to achieve safe military disposition of, this material. They continued their battle earlier this year amid reports that the administration of President George W. Bush was going to halt the efforts to immobilize plutonium in nuclear waste. Immobilization is greatly preferable to recycling plutonium into mixed-oxide fuel (a mixture of uranium and plutonium oxides) for use in commercial power plants. Leventhal and the institute have been alerting the world to the risks of using mixed-oxide fuel, making clear to all that this fuel, if stolen by terrorists, could be used to make bombs, and that a radiological release from a reactor accident involving this fuel would pose a great cancer threat to the public.

My twenty-five years in Congress, which have witnessed the birth and growth of the Nuclear Control Institute, have focused on the connection between nuclear power and nuclear arms. From the attempt by President Jimmy Carter in 1980 to ship nuclear fuel to India's Tarapur plant to the present possibility of supplying nuclear materials and technology to the Democratic People's Republic of Korea (North Korea), I have battled against the weakening of the nonproliferation regime, which threatened to propagate the nuclear genie around the world. Leventhal and the Nuclear Control Institute have been right there beside me, providing their invaluable expertise and support.

But nuclear proliferation has a way of evolving into even more insidious forms. When I wrote *Nuclear Peril: The Politics of Proliferation* in 1982, the focus was on nation states, such as India, Pakistan, and Iraq, racing to enter the nuclear weapons club through the nuclear power door. Today, we must also be prepared for subnational groups and terrorist cells that may attempt to acquire and use nuclear weapons, or even radiological weapons, wherein radioactive materials are used with conventional explosives to make a terror bomb.

Furthermore, the tragic events of September 11, 2001, speak to even more chilling scenarios. For while the terrorist attacks of September 11 did not involve nuclear facilities or materials, they demonstrated that dedicated individuals are not reluctant to transform civilian technology into weapons of terror. Imagine, for example, the tragedy that would have resulted had the targets of the hijacked airplanes been nuclear power plants. The saddest irony is that the nuclear industry, in attempting to avoid real improvements in security over the last decade, has failed to recognize what the implications would be for public confidence in their industry if one—*just one*—plant's security was compromised.

Even before September 11, the Nuclear Control Institute and I had worked together to pressure the Nuclear Regulatory Commission to increase protection against terrorist threats at civilian nuclear power plants. Since September 11, we have continued our collaboration with renewed determination. We have called for National Guard protection at nuclear power plants. We have called for distribution of potassium iodide to communities around nuclear power plants to protect the thyroid against radioactive byproducts in the event of a release. We have called on the Nuclear Regulatory Commission to upgrade the design basis threat—the hypothetical scenarios of attacks against which plants must be defended. We have called for the continuation and augmentation of the Operational Safeguards Response Evaluation program. I know that the Nuclear Control Institute will be there as we continue the fight to protect the people of our country from potential nuclear tragedies. Complacency will be no friend of ours in this struggle.

The contents of this volume were written well in advance of the clarion call of September 11. But in this new era they are even more valuable, giving us insight into the many dangers that we face and forcing us to review our reliance on nuclear power. There is a renewed urgency to understand what the authors have written. The Nuclear Control Institute does us a great service by collecting and presenting their words in print.

I look forward to more years of fruitful battle alongside the Nuclear Control Institute. And I extend a warm salute to Leventhal and the institute as they continue their vital work to keep this country, indeed the world, safe from the potential dangers of nuclear materials.

U.S. Representative Edward J. Markey

PREFACE

I started the Nuclear Control Institute in 1981 after landing a $7,500 contribution from an anonymous member of the Rockefeller family. Wade Greene, the Rockefeller family's representative, called it a "stimulative grant" to encourage giving by other foundations. I had just lost my job on Capitol Hill, when the majority of the Senate switched to the party other than the one my boss and subcommittee chairman, Gary Hart, belonged to. Thus, I wasted no time and applied the Rockefeller check to renting a desk in the corridor of a small law firm in a downtown Washington, D.C., townhouse. With the desk came a posh conference room suitable for holding meetings on plutonium and proliferation. The Nuclear Control Institute was born, and thus began a two-decade process of research and advocacy that culminated in the conference on which this book is based.

I owe a special debt to Wade Greene and the family he represents not only for the institute's initial grant but for sustained financial support and encouragement over the years, especially their establishment of the Nuclear Control Institute Sustaining Fund that provided the money for convening the institute's twentieth anniversary conference.

In the early days, "NCI" stood for The Nuclear Club Inc. Our name was surely too clever by five-eighths. But we used it anyway in a full-page *New York Times* advertisement on Sunday, June 21, 1981, to launch the fledgling organization. The ad posed the question, "Will Tomorrow's Terrorist Have an Atom Bomb?" It is a question, unfortunately, that is especially relevant today—as is the answer. The institute's name has changed, but our mission—to prevent the further spread of nuclear weapons to nations or to groups—remains the same.

The ad's creator was Julian Koenig, an original member (and still a member) of the institute's board of directors. He is a Madison Avenue legend, now retired, whose credits included Volkswagen's original

"Think Small" campaign and the naming of Earth Day. At first Mr. Koenig expressed reluctance about joining our board, but I assured him that the institute would have to solve the plutonium problem within five years or he and I probably would not survive to talk about it.

I was wrong on both counts. We have not solved the problem. We are still around to talk about it. To paraphrase Faulkner, the Nuclear Control Institute has endured, if not prevailed. We are all still here to explore the role of nuclear power, plutonium, and other associated proliferation risks—the purpose of this book.

The institute's endurance has been made possible by a number of foundation officers, especially those who were willing to make an early bet on the first organization to be devoted exclusively to nuclear nonproliferation—Hilary Palmer of the Rockefeller Brothers Fund, Sally Lilienthal of the Ploughshares Foundation, John Redick of the W. Alton Jones Foundation, Fritz Mosher of the Carnegie Corporation of New York, Ruth Adams of the MacArthur Foundation, and Enid Schoettle of the Ford Foundation. In more recent years we are indebted to the financial and moral support of George Perkovich of the W. Alton Jones Foundation, Kennette Benedict of the MacArthur Foundation, Coco Eiseman of the Prospect Hill Foundation, Sandra Silverman of the Scherman Foundation, and Nancy Stockford of the John Merck Fund.

I also wish to acknowledge the invaluable support and advice received from those who have served at various times on the Nuclear Control Institute's board of directors, a number of whom have made a distinguished professional contribution to nuclear nonproliferation in their own careers: the late Rear Admiral Thomas D. Davies, the late Barbara Tuchman, Theodore B. Taylor, Peter A. Bradford, Roger Richter, Sharon Tanzer, Julian Koenig, Victor Gilinsky, Denis Hayes, and David Cohen.

The work of the Nuclear Control Institute has its roots on Capitol Hill, where I pursued the nuclear proliferation problem before founding the institute. It is important to acknowledge the contribution of the senators for whom I worked when the spread of nuclear weapons was still a relatively new issue—the late Senator Abraham Ribicoff (D-Conn.), Senator Charles Percy (R-Ill.), and Senator John Glenn (D-Ohio). Their involvement reflected a special commitment and act of political courage. I originally came to Washington in 1969 to work as press secretary for the late Senator Jacob K. Javits (R-N.Y.), and although this was prior to my immersion in nuclear issues, the influence on my later work of this brilliant, indefatigable senator was profound.

Other members of Congress to whom the Nuclear Control Institute owes a debt of gratitude for their substantial contributions over the years toward reducing the dangers of nuclear proliferation are the late Senator Alan Cranston (D-Calif.), the late Representative Jonathan Bingham (D-N.Y.), Representative Edward J. Markey (D-Mass.), who prepared the foreword to this book, and former Representatives Richard Ottinger (D-N.Y.) and Howard Wolpe (D-Mich.). On Capitol Hill, initiatives on nuclear nonproliferation, and especially my own early efforts, were immensely benefited by the research and analyses of the late Warren H. Donnelly, senior nonproliferation specialist at the Congressional Research Service. It is also important to acknowledge the early contributions of Richard A. Wegman, former staff director and general counsel of the Senate Government Operations Committee, Leonard Bickwit, former legislative assistant to Senator Glenn, and Constance Evans, former legislative assistant to Senator Percy, in undertaking the original investigation of U.S. nuclear export policies that led to the enactment of the Nuclear Non-Proliferation Act of 1978.

The record of accomplishment of the Nuclear Control Institute would not be possible without the diligence, tenacity, and plain hard work of its excellent staff, present and past. I offer my special appreciation to Sharon Tanzer, the institute's vice-president, whose outstanding service and wise counsel extend over the full twenty years. I offer my heartfelt thanks to Tom Clements, executive director; Edwin Lyman, scientific director; Steven Dolley, research director and webmaster; and past staffers John Buell, Daniel Horner, Milton Hoenig, Deborah Holland, and Laura Worby. Alan Kuperman, a former staffer and now the institute's senior policy adviser, deserves special acknowledgement for the singular role he has played on the institute's behalf in helping to achieve the virtual elimination of U.S.-origin bomb-grade uranium from world commerce. In addition, our two outside counsels, Eldon V. C. Greenberg and my former Capitol Hill colleague, Mr. Wegman, both partners in the law firm of Garvey, Schubert, and Bayer, have made invaluable contributions to the effectiveness of the institute's litigation and strategic initiatives. The institute's technical analyses have been enriched by the remarkable contributions of the late J. Carson Mark, former director of the theoretical division of Los Alamos National Laboratory, and of Marvin Miller of the Massachusetts Institute of Technology. I am also particularly indebted to our brilliant website designer, Bojan Dobrecevic in Slovenia, for his tireless efforts to give the institute a presence on the worldwide web (www.nci.org) that is attractive and lively, as well as informative. In addition, Morton Cohen's service as the institute's financial adviser and Stephen M. Johnson's role as public relations adviser are warmly appreciated.

Of course, the work of the Nuclear Control Institute is sharpened by those with whom we interact in addressing nuclear proliferation problems. I want to acknowledge in particular the following contributions: Armando Travelli and his technical team at Argonne National Laboratory in leading the remarkably effective international program to convert research reactors to fuel unsuitable for weapons; Daniel Hirsch, president of the Committee to Bridge the Gap, with whom we have collaborated for almost two decades on upgrading the physical protection of nuclear power plants; Ambassador Robert L. Gallucci for his occasional, incisive counsel and for his years of courageous and effective government service; Randy Rydell, who as an aide to Senator Glenn made invaluable contributions to the cause of nonproliferation; Lawrence Scheinman, a scholarly defender of the nonproliferation role of international organizations; Shuzaburo Takeda of Tokai University, who facilitated my first trip to Japan and made possible the institute's work there; Aileen Mihoko Smith, president of Green Action Kyoto, and the late Jinzaburo Takagi, founder and president of the Citizens' Nuclear Information Center, whose efforts, aided by Mara Takubo of Gensuiken, to bring a halt to the Japanese plutonium program have been heroic; and Damon Moglen and Shaun Burnie of Greenpeace International, whose actions on behalf of nonproliferation stand out in a field too dominated by talk.

I would be remiss if I did not also acknowledge the worthy adversaries whose efforts on behalf of nuclear industry interests did not prevent them from engaging in meaningful discourse with me and my colleagues at the institute. To Bertram Wolfe, Harold Bengelsdorf, Fred McGoldrick, Howard Shapar, Roger Gale, and Daniel Poneman I express my appreciation for their willingness to avoid the "dialogue of the deaf" that often afflicts this field, and to engage the opposition, often with good humor.

Finally, to the extent this book presents a cogent, incisive discussion of complex energy and nonproliferation issues, I am deeply indebted to Drs. Miller and Lyman for suggesting the themes of the need for nuclear power in relation to the alternatives, and of nuclear power's irreducible proliferation risk, respectively. My co-editors, Ms. Tanzer and Mr. Dolley, contributed greatly to the planning and organizing of the book and of the conference that preceded it. To the extent the book explores complex issues in a highly readable way that makes sense to general readers, I am deeply indebted to our talented and persevering editor, Helen Whitney Watriss. In addition, Christina Davidson, assistant editor at Brassey's, Inc., made important contributions to the concept and the style of the book. Ultimately the contribution this book makes to a better understanding of the risks and benefits of nuclear power in an increasingly energy- and security-challenged world rests upon the considerable expertise and the wisdom of the conference participants and contributing essayists. Their time and their commitment are deeply appreciated.

PAUL L. LEVENTHAL
Washington, D.C.
December 8, 2001

1

Introduction: Nuclear Power without Proliferation?

Paul L. Leventhal

The essays in this book are based on the presentations made at a conference held in Washington, D.C., on April 9, 2001, to mark the twentieth anniversary of the Nuclear Control Institute. At the time of the conference, the administration of President George W. Bush was launching a new energy policy to respond to a perceived national energy crisis, which had been prompted by the sudden appearance of electricity shortages, blackouts, and soaring natural gas prices in California. A rebirth of nuclear power was high on the administration's list of long-term solutions to this crisis.

Marvin Miller, a nuclear physicist from the Massachusetts Institute of Technology and a longtime adviser to the Nuclear Control Institute, suggested to me that our institute consider using the conference to examine the question of whether nuclear power reactors were needed to respond to electricity shortages and whether the proliferation risks associated with nuclear energy made the search for alternative means of generating electricity an imperative. Miller's suggestion gave rise to the name of our conference, "Nuclear Power and the Spread of Nuclear Weapons: Can We Have One without the Other?"

As we were completing the book manuscript nearly eight months after the conference, the crisis environment in Washington had been entirely transformed. Gone, at least for the time being, was the energy crisis—thanks to the restoration of a balance between electricity supply and demand, as well as a sharp drop in natural gas prices, both in California and the nation at large. Replacing the energy crisis was the nation's mobilization for war against terrorism in response to the surprise attacks of September 11, 2001, on the World Trade Center and the Pentagon.

This book's examination of nuclear power—the need for it and the risks associated with it—is highly relevant to the new threat environment posed by highly sophisticated, well-coordinated, and suicidal terrorists dedicated to mass killing. Of immediate concern are the adequacy of the protection of nuclear power plants against attacks that could cause widespread and deadly radioactive contamination, and the adequacy of the protection of plutonium separated from the spent fuel of these plants against theft for use in atomic bombs. A longer term concern is the adequacy of the worldwide nuclear nonproliferation regime and, in particular, its ability to ensure that nuclear power and research programs do not serve as a cover for the development of nuclear weapons in additional nations, especially those that support terrorism. These issues have long been of concern to the Nuclear Control Institute, and they are addressed in this book.

Over two decades, the institute's principal focus has been the nuclear proliferation and terrorism risks associated with nuclear power. In this book we also examine a policy area into which we have not ventured before. We look at nuclear power in a broader context: Does the world need nuclear power? How essential is it? What are the possible alternatives to it? The contributions in this book are by world-class experts who present a full range of viewpoints. Their original presentations at our conference and their interchanges with each other and with an expert audience are recorded in the rapporteur's summary of the conference, prepared by Steven Dolley, which follows in the next chapter. The inclusion of opposing viewpoints at the conference generated heat, but also light.

Although some in industry and bureaucracy have concluded that the institute's opposition to civilian use of plutonium and the other nuclear weapon material, highly enriched uranium, means that it opposes nuclear power, it is in fact not an antinuclear organization. The institute has maintained a policy of neutrality on nuclear power and steered clear of efforts to shut the industry down. It is antiplutonium and highly enriched uranium, not antinuclear.

The push by industry and by the Bush administration to revive and expand nuclear power in response to concerns about electricity shortages and global warming prompted this examination of nuclear power and the alternatives. Vice President Richard B. Cheney, as reported in the *Washington Post* of April 9, 2001, said, "We need to build sixty-five new power plants for the next twenty years, and my own view is that some of those ought to be nuclear, and that's the environmentally sound way to go." The present direction of the nuclear industry and the Bush administration win strong support in the essays by Richard Rhodes and Bertram Wolfe.

The Nuclear Control Institute believes that such an initiative should not go forward without first examining whether there is an irreducible risk of proliferation associated with nuclear power, and whether this risk is serious enough to compel a change in current commitments to nuclear power. If the nuclear industry refuses to end its love affair with plutonium—especially now that experts widely acknowledge it is not an essential fuel because of the abundance of cheap, nonweapon-usable uranium—then the world may well be better off without nuclear power. In that case policymakers must look to alternative sources of energy and to energy conservation and efficiency measures. These are the issues that Richard Rhodes, Amory Lovins, Richard Garwin, and Robert Williams debate in their essays.

Even if industry does give up using plutonium, there are still severe proliferation dangers associated with the prospect for potentially limitless sources of uranium and for cheap, efficient technologies that could be used to enrich it to bomb-grade. Garwin and Harold Feiveson include discussion of the uranium issue in their essays.

By way of context for the ensuing discussions, it is important to highlight some of the Nuclear Control Institute's concerns about the proliferation and security risks of nuclear power and how these risks are being addressed. A key concern is the central role of *fissile materials* as the driving force behind proliferation. Although any decision to go nuclear is a political one, its execution is technical. It is impossible to build nuclear weapons without plutonium or highly enriched uranium. It should be straightforward, therefore, that the nuclear power industry imposes a menace on the world if it insists on utilizing these explosive nuclear fuels when it is possible to run nuclear power and research reactors without them.

As Robert Gallucci, Edward Markey, Marvin Miller, and William Lanouette discuss in their essays, history shows that nuclear power programs *have* provided cover for actual or attempted weapon-making in a number of countries. In each case the quintessential elements of those programs were closing the fuel cycle to extract plutonium, enriching uranium to weapon-grade, or importing weapon-grade uranium to run research reactors. Zachary Davis, George Perkovich, and Lawrence Scheinman discuss India and Pakistan's use of civilian nuclear programs to develop nuclear weapons, as well as concerns that Iran is headed in the same direction, and the implications of these developments for the global nuclear nonproliferation regime. I provide a response to their presentations.

The objective of the congressional nonproliferation initiatives of the 1970s and of the administrations of Presidents Gerald R. Ford and Jimmy Carter was to restrict and eliminate use of these fuels. These initiatives ran into political trouble, however, because America's European and Japanese allies were fiercely opposed and refused to follow the U.S. example. Today, the plutonium and breeder programs in those countries are in desperate financial straits, and the United States now has an opportunity to reopen these issues and seek cooperative approaches to the disposal of excess fissile materials other than introducing them as fuels.

In a recent analysis of prospects for the nuclear industry, the pro-plutonium British Nuclear Industrial Forum stated: "Proliferation is a major issue in the nuclear fuel cycle. Nuclear power may become more acceptable to the public if reprocessing is shut down."[1] Clearly the plutonium program in Great Britain, as in France, Germany, and Japan, is encountering significant difficulties. British Nuclear Fuels Ltd., the government-owned fuel cycle company, agreed, in the course of a stakeholders' dialogue on its plutonium program, to undertake a formal assessment of the immobilization of Britain's 60-plus-ton stockpile of civilian plutonium as an alternative to fabricating it into plutonium-uranium, mixed-oxide fuel. At the same time, British Nuclear Fuels Ltd. is pressing ahead, over the objections of its Irish and Nordic neighbors, with plans to operate a new plant for fabricating mixed-oxide fuel and to continue operation of a large reprocessing plant for separating plutonium from spent fuel, despite an enormous plutonium surplus.

Britain's plutonium dilemma, as well as the setbacks faced by the plutonium industry elsewhere, present an opportunity for the United States to revisit the plutonium component of its nonproliferation policy. But "transparency" and "gradualism" still dominate U.S. policy. Achieving transparency of the world's civilian plutonium stockpiles is, however, no substitute for getting rid of them, while gradualism can be an excuse for not doing anything effective. The rapid growth in world stocks of surplus plutonium illustrates this point because it is largely attributable to the failure by the United States to exercise its legal prerogative to say no to countries that wanted to separate plutonium from the nuclear power plant fuel we provided them.

The growth in plutonium stocks has not been as rapid as the Nuclear Control Institute projected in 1983 when it commissioned David Albright to do his first study of this subject.[2] At that time the institute projected 600 tons of civilian plutonium separated from spent fuel by the year 2000. Today, because of large-scale cancellations of orders for new nuclear power and fuel-cycle plants, and because of the demise of the breeder reactor, about 300 metric tons[3] of plutonium have been separated, of which only about one third has been irradiated as mixed-oxide fuel. That leaves about 200 tons of unirradiated civilian plutonium—still an awesome figure that approximates the amount of military plutonium in the world.

By way of contrast, it should be noted that stocks of civilian highly enriched uranium exported by the United States have declined dramatically as a result of the International Reduced Enrichment for Research and Test Reactors program run by the U.S. Argonne National Laboratory, with modest support from the executive branch. In this case there is a law, the Schumer amendment, that takes a sanctions approach and bars exports of highly enriched uranium except to research reactors whose operators have agreed to convert to high-density, low-enriched uranium that cannot be used in bombs. The result is that highly enriched uranium exports by the United States are now virtually nil, limited to relatively small amounts that support the continued operation of reactors as they prepare to convert.

Plutonium is a different story. The executive branch has circumvented provisions in the Nuclear Non–Proliferation Act of 1978 that are intended to restrict commerce in plutonium derived from U.S.–supplied nuclear fuel. The Bush administration's call for a major expansion of nuclear power, in combination with reconsideration of reprocessing and use of plutonium as reactor fuel, is an invitation to catastrophe—a twenty-first century siren's song. It is especially worrisome because it is misinformed.

A push for nuclear power and widespread use of plutonium is not the way to meet urgent energy needs. New plants cannot be brought on line fast enough to offset immediate electricity shortages and will do little to reduce emissions of greenhouse gases. As Williams discusses in his essay, two thirds of the emissions of carbon dioxide are from transportation and other sources not related to power generation. If nuclear were used to replace coal plants on a global scale, there would have to be a tenfold increase in nuclear capacity to three thousand 1,000-megawatt (electric) plants. Carbon emissions would be reduced by only 20 percent, but millions of kilograms of separated plutonium—equivalent to hundreds of thousands of bombs—would have to enter commerce each year to support the breeder reactors that no doubt would be required to sustain such a vast commitment to nuclear power. Furthermore, Lovins asserts in his essay that energy efficiency measures, using the best existing technologies, could save the United States three quarters of its electricity use—equivalent to four times current nuclear output and costing less to install than current nuclear operating costs.

The Bush energy plan also presses for pyroprocessing, a next-generation variation of reprocessing, and for accelerator transmutation of plutonium and other long-lived radioactive products in spent fuel. Both approaches to reprocessing are uneconomic and dangerous. The plan fails to mention the enormous cost projected for establishing a national pyroprocessing and transmutation system. The U.S. Department of Energy estimated in 1999 that this program would cost $280 billion and take 117 years to complete—very likely a gross underestimate given the department's average cost overrun of 500 percent on large capital projects. Transmutation, which involves pyroprocessing, reprocessing, and subcritical nuclear reactors, is highly problematic and in any event does not eliminate the need for a final waste repository. James Hassberger and Edwin Lyman debate these matters in their essays.

The Bush energy plan cites the reprocessing and waste management experience of Britain, France, and Japan as an example for the United States to follow. In fact, neither Britain, France, nor Japan has a long-term plan to dispose of the large volumes of high-level waste associated with reprocessing. They are currently struggling not only with what to do with reprocessing waste, but also with their rapidly growing stockpiles of unwanted, uneconomical, weapon-capable plutonium.

The French national utility, Electricité de France, recently admitted that reactor fuel made with separated plutonium is three to four times more expensive than the conventional fuel made with low-enriched uranium.[4] The British plutonium program has proved an economic and technological disaster, with a stockpile now approaching some 70 metric tons of separated plutonium and no domestic utilities willing or able to use it.

The Bush administration is still pursuing a program to dispose of excess weapons plutonium by turning it into mixed-oxide fuel for use in U.S. and Russian power reactors. This is the most dangerous course, especially in Russia, where security and safety measures fall far below western standards. In the United States, this program is breathing new life into a moribund plutonium industry and giving hope to those like Senator Pete V. Domenici (R–N.M.) who still advocate eventual separation of plutonium from commercial spent fuel. To make matters worse, the Bush administration has killed funding for the faster, cheaper, safer course for disposing of military plutonium—immobilization of the plutonium by combining it with highly radioactive waste. This is the so-called "second track" begun under President William J. Clinton. Surplus weapons plutonium in the United States and Russia, as well as commercial plutonium worldwide, should be immobilized in existing highly radioactive, self-protecting nuclear waste, not used in reactors. Indeed, as I discuss elsewhere in my essay, immobilization could provide common ground for nuclear-weapon states and non-nuclear-weapon states to establish a symmetrical approach for getting rid of excess plutonium stocks in a way that upholds both the Nuclear Non-Proliferation Treaty and nuclear disarmament objectives.

It is important to note *the pivotal role of Japan* in perpetuating plutonium commerce. The Nuclear Control Institute has been focusing significant attention on Japan's plutonium program, and sometimes we are criticized for doing so. Why is the institute so concerned about plutonium in Japan, given that country's adherence to the Nuclear Non-Proliferation Treaty and the safeguards of the International Atomic Energy Agency?

The answer is that Japan strongly resisted early U.S. efforts to avoid commercial use of plutonium and is now the lynchpin of world plutonium commerce. With Germany getting out of the plutonium business, Japan is the most important customer of the European reprocessing and mixed-oxide fuel industries. Without Japan these industries might well be forced to shut down.

The Japanese plutonium program has been losing domestic public acceptance because of a succession of nuclear accidents in Japan and a scandal that developed when workers at British Nuclear Fuels Ltd. deliberately falsified the quality-control data for the mixed-oxide fuel it was shipping to Japan for use in light-water reactors. Elsewhere in the East Asia region there is considerable suspicion as to why Japan wants to accumulate so much weapon-usable plutonium, given the clear alternative of low-enriched uranium fuel. In a detailed economic analysis the Nuclear Control Institute has pointed out that Japan could ensure its energy security by building a strategic reserve of non-weapon-capable uranium at a fraction of the cost of its plutonium and breeder programs.[5]

The Nuclear Control Institute also regards Japan as a special case because, of all the civil plutonium-consuming countries, it refuses to acknowledge the weapon utility of reactor-grade plutonium despite many briefings on the subject by the U.S. government. The late J. Carson Mark, former head of weapons design at Los Alamos National Laboratory, was commissioned by the Nuclear Control Institute to do an analysis of the weapons utility of reactor-grade plutonium.[6] The results eventually persuaded the International Atomic Energy

Agency to acknowledge that reactor-grade plutonium was suitable for weapons.[7] Unfortunately the Japanese government and industry still refuse to do so.

The Japanese plutonium program also has prompted strong protests from many nations because of alarm over the regular transports of mixed-oxide fuel and highly radioactive reprocessing waste that now pass close to their coastlines along routes from Europe to Japan. Japan has not been responsive to their safety and security concerns or to their demands for environmental impact assessments, advance consultation on emergency planning, guarantees of salvage of lost cargoes, and indemnification against the catastrophic consequences of accidents or attacks. The result is that the Japanese plutonium program is mired in controversy, internationally as well as domestically.

The Nuclear Control Institute believes that Japan should be regarded as a special case and a special concern. If Japan should eventually decide against further use of plutonium fuel and the European plutonium industry collapsed as a result, it might then be possible to build an international consensus to eliminate commerce in plutonium as well as bomb-grade uranium. Japan and the other big plutonium-producing and consuming countries *are* critically important to nuclear nonproliferation because they set an example and standard for the rest of the world.

It is also important, especially in the wake of the terrorist attacks of September 11, to highlight the Nuclear Control Institute's concerns about the possibility of *reactors being used as radiological weapons*. The risk of sabotage of nuclear power plants is not just a problem in Russia, where security is widely regarded as lax. In the United States, half the nuclear power plants have failed to repel mock attacks—so-called force-on-force exercises supervised by the Nuclear Regulatory Commission. Despite these failures the commission has refused to take enforcement action, is even weakening the rules of the game, and is preparing to transfer supervision of the exercises to nuclear plant operators—all in response to industry complaints. When the mock attackers reach and "destroy" a complete set of redundant core-cooling systems, which has been the result in nearly half of the cases, the agency even refuses to acknowledge officially the "pass-fail" nature of the exercises. Indeed, the grading system for these exercises could well be called "pass-melt."

The sabotage of nuclear power plants may be the greatest domestic vulnerability in the United States today. The mock adversary in these exercises has been comprised of far fewer attackers than were encountered on September 11. The new threat against which plants must be defended, as made manifest on September 11, is at least nineteen terrorists attacking from four different directions. Against this threat, nuclear plants do not now have adequate protection. In the past industry operators have been unprepared to pay the cost of achieving adequate protection, and the Nuclear Regulatory Commission has been ill-disposed to require it. It is now not clear that conventional, private guard forces can assure the security of nuclear power plants against attack and sabotage. Military protection is required. This issue warrants a hard look.

This issue also raises the larger question of the *adequacy of nuclear regulation today*. It is essential to maintain strong, independent nuclear regulation, free of undue industry influence. The Energy Reorganization Act of 1974 "fissioned" the Atomic Energy Commission into two separate regulatory and promotional agencies and thereby transformed the weak regulatory division of the Atomic Energy Commission into a strong, independent Nuclear Regulatory Commission. Today, however, the commission is looking more and more like the old regulatory division of the Atomic Energy Commission, subject to undue influence by industry and, in particular, industry's powerful friends on Capitol Hill. This matter also deserves close scrutiny.

Over the past two decades at the Nuclear Control Institute and in my earlier work on Capitol Hill, I have been influenced by two leading nuclear contrarians. One was the late David Lilienthal, who served as both the first head of the Tennessee Valley Authority and the first chairman of the Atomic Energy Commission. His Senate testimony in 1976 in opposition to U.S. nuclear exports and in support of nonproliferation legislation caused a furor among his former colleagues. He once said to me, "If we assume nuclear proliferation to be inevitable, of course it will be." That statement still makes a lot of sense today.

Theodore Taylor, America's most creative fission bomb designer and a member of the institute's board of directors, also made a concise and similarly compelling point to me: "Nuclear is different," he said. To illustrate his point, Taylor noted that the bomb that destroyed Nagasaki set off an instant of explosive energy equivalent to a pile of dynamite as big as the White House, all contained in a sphere of plutonium no bigger than a baseball. That first-generation bomb is a technological feat now within the grasp of terrorists or radical states that manage to get their hands on the material.[8]

The issues raised by Lilienthal and Taylor, and the questions examined in this book, suggest that ultimately a test of reasonableness must be applied. Is it reasonable to assume that, over time, millions of kilograms of plutonium can be sequestered down to the less than 8 kilograms needed for an atomic bomb? This question must be answered before giving any further comfort and support to an industry that remains officially committed to utilizing plutonium as a fuel—and surely before supporting an extension and expansion of that industry in response to electricity-supply shortages and to global warming.

In closing, it is useful to recall the point made by one of the Nuclear Control Institute's original board members, the late historian Barbara Tuchman. In one of her books she gave a sobering description of the "march of folly" that drives nations to destruction. The phenomenon she identified, one repeated throughout recorded history, is a "perverse persistence in a policy demonstrably unworkable or counterproductive."[9] To qualify as folly, she said, a policy "must have been perceived as counter-productive in its own time, not merely by hindsight . . . [and] a feasible alternative course of action must have been available."[10]

In that context, it is vital to explore nuclear power and the alternatives.

2

Rapporteur's Summary of the Nuclear Control Institute Twentieth Anniversary Conference

Steven Dolley

The twentieth anniversary conference of the Nuclear Control Institute, "Nuclear Power and the Spread of Nuclear Weapons: Can We Have One without the Other?," took place on April 9, 2001, at the Carnegie Endowment for International Peace in Washington, D.C. The names and biographies of the sixteen presenters can be found in the biographical notes on the contributors at the back of this book. In addition, more than a hundred people attended the conference and participated in discussions that followed the presentations.[1] This chapter provides a brief summary of each presentation and the highlights of the discussions.

Introduction

Nuclear Power without Nuclear Proliferation? Paul L. Leventhal, Nuclear Control Institute

There is currently much renewed interest in nuclear power, as evidenced by the plans of U.S. electric utilities to construct next-generation reactors and Vice President Richard B. Cheney's April 8, 2001, statement that the United States will need to build sixty-five new power plants annually and at least some should be nuclear. The conference examines the future of nuclear power by posing some basic questions:

1. How essential is nuclear power?
2. Are there viable alternatives?
3. Is there an irreducible proliferation risk from nuclear power that makes the alternatives imperative?

The role of fissile materials is central to the proliferation of nuclear weapons. Incentives for nations to go nuclear may be primarily political, but the question of capability is technical. It is not possible to build nuclear weapons without either plutonium or highly enriched uranium. Nuclear power presents a menace to the world if the industry continues to insist on producing and utilizing these weapon-usable fuels when power and research reactors can be run without them.

Nuclear power programs have also provided cover for weapon development in a number of countries. In the 1970s the restriction of weapon-usable fuels was the objective of congressional nonproliferation initiatives (including the Nuclear Non-Proliferation Act of 1978), as well as of the nonproliferation policies of the administrations of Presidents Gerald R. Ford and Jimmy Carter. Western Europe and Japan strongly opposed those initiatives and policies. The United States canceled its commercial plutonium and fast breeder reactor programs, but other nations did not follow its lead. Today the heavily subsidized plutonium and fast breeder reactor programs around the world are in desperate straits because of accidents, poor economics, and the risks of proliferation.

To date about 300 metric tons[2] of plutonium have been separated in civilian programs, of which only about one third has been irradiated as plutonium-uranium mixed-oxide fuel. This leaves about 200 tons of unirradiated plutonium in the civilian sector, an amount that approximates all the military plutonium in the world. Separation of plutonium from U.S.-supplied reactor fuel continues to increase each year—the consequence of U.S. policy dominated by considerations of gradualism and transparency rather than action to halt the process. Yet U.S. exports of highly enriched uranium declined significantly and now are virtually nil, the result largely of the U.S.-sponsored conversion of research reactors to low-enriched fuels under the auspices of the Reduced Enrichment for Research and Test Reactors program.

A focus on Japan's nuclear program is justified because Japan is the biggest customer for the Western European reprocessing and mixed-oxide fuel industries. Suspicion about Japan's accumulation of separated plutonium is growing in East Asia. Japan does not acknowledge the weapons utility of plutonium separated from nuclear power reactor fuel (so-called "reactor-grade plutonium") despite receiving U.S. government briefings to clarify the matter. Even the International Atomic Energy Agency acknowledges that all isotopic mixes of plutonium (except pure uranium-238) can be used to make bombs. The behavior of the big countries counts in nonproliferation policy: They set the example and standard for the rest of the world.

Another risk is sabotage. Nearly one half of all U.S. nuclear power plants have failed force-on-force exercises involving mock terrorist attacks, yet the U.S. Nuclear Regulatory Commission refuses to take enforcement action and is weakening the rules. Radiological sabotage may prove to be the greatest danger of nuclear power. Indeed, the Nuclear Regulatory Commission is looking more and more like the regulatory division of the old Atomic Energy Commission, in that the nuclear industry once again exercises undue influence upon the regulatory process.

Former Atomic Energy chairman David Lilienthal once said, "If we assume nuclear proliferation to be inevitable, of course it will be." We must ask if safeguards are up to the task of protecting the public from a reinvigorated nuclear industry that refuses to give up its love affair with plutonium, and whether we can control the world's plutonium stocks down to the 8 kilograms or less needed to make a bomb.

Panel: How Essential Is Nuclear Power?

Nuclear Power and Proliferation, Richard Rhodes, Author

Electricity's share of energy is increasing rapidly worldwide. Energy policy is not a zero-sum game. We need to encourage conservation, develop a variety of energy sources, and decrease coal combustion. Nuclear power provides 20 percent of U.S. electricity and increased its output by over 20 percent from 1990 to 2000 without building any new plants. By contrast, renewable energy now provides about 2 percent of U.S. electricity.

Nuclear power was once heavily subsidized, but by 1997 the federal research and development budget was thousands of times greater (per unit of energy produced) for wind and solar than for coal and nuclear. Energy saved by subsidized energy conservation is about twice as expensive as newly generated power.

U.S. nuclear power's average capacity factor improved from 58 percent in 1980 to 90 percent today. By contrast, a Wisconsin wind farm recently averaged only 24 percent. Both wind and solar suffer from low capacity that, unlike nuclear power, cannot be greatly improved.

Most U.S. air pollution and emissions of carbon dioxide come from cheap but deadly coal combustion. Coal generation of electricity releases a hundred times more radiation per megawatt than does nuclear power. A Harvard study concluded that air pollution from coal combustion kills fifteen thousand people per year in the United States; other studies estimate up to thirty thousand deaths annually. Nuclear power could replace coal-fired generation, avoiding millions of tons of emissions of pollution.

Nuclear waste disposal is a political, not technical, problem. Experts from twenty nations agreed in 1985 that nuclear waste could be disposed of safely using available technology. Well over 99 percent of the toxicity from nuclear waste decays after five hundred years.

Some nuclear power opponents want to return to an agrarian, crafts-based lifestyle. The green movement has been called an effort to overturn industrial society; nuclear power threatens that goal by offering an abundant, stable energy source. There is no such thing as a nonpolluting energy source. To achieve the optimal worldwide level of electricity consumption, we would need to generate 50–100 percent more electricity. Without sufficient energy, the world faces structural violence and military conflict. Two billion people in the world have no electricity; to deny them energy is elitist and immoral.

Proliferation is also a political, rather than technical, problem. No nation has gone nuclear using plutonium from power reactors, and alternate means to acquire plutonium for weapons are superior. No nation that ratified the Nuclear Non-Proliferation Treaty has become a nuclear weapon state. Proliferation does not increase with more nuclear power. Indeed, eliminating nuclear power might encourage proliferation by increasing structural violence caused by scarcity.

The 1946 Acheson-Lilienthal report proposed international control of nuclear power and fissile materials. Plutonium's risks are best averted by burning it as fuel in reactors, including fast reactors. Scientists at Los Alamos National Laboratory have proposed a global regime for the management of nuclear materials based on secure retrievable storage at internationally controlled sites, an approach that could later go beyond storage to develop advanced, proliferation-resistant reactors.

The ultimate driver of proliferation is the arsenals of nuclear-weapon states, not nuclear power. Abolition of all nuclear weapons is the only solution. Nuclear power today is alive and well and remains an important part of the solution to global energy problems.

Why Nuclear Power's Failure in the Marketplace Is Irreversible (Fortunately for Nonproliferation and Climate Protection), Amory B. Lovins, Rocky Mountain Institute

Nuclear power is dying a lingering death because of market forces. Renewable energy alternatives are flourishing. All industry literature shows renewables are increasing their share of energy supply. We can develop efficient end-use of energy, gas, and wind power—any one of which could make nuclear power unnecessary and uneconomic.

For remote electricity generation, one must include the cost of delivery. Current nuclear-generated electricity costs 10–15 cents per kilowatt-hour delivered, more than double the delivered cost of combined-cycle gas-generated electricity or even the current generation of wind-power electricity.

Efficient end-use has become the nation's largest energy supply, providing over five times more energy than does domestic oil and over twelve times more than Persian Gulf imports. The United States could save up to an additional 75 percent of electrical use at a cost below that of short-run marginal supply. That amounts to more than four times the present output of U.S. nuclear power, at a price much cheaper than its operating cost alone. Historically, electricity from efficient end-use costs about two cents per kilowatt-hour in the United States, and it will become even cheaper. Each six cents spent on one nuclear kilowatt-hour could have purchased *two* kilowatt-hours from efficient end-use. Expanding returns allow us to "tunnel through the cost barrier" with a new design mentality, achieving greater than 100 percent return on investment, often within one year. A Chicago office tower was retrofitted to save three quarters of its energy use at no greater cost than a standard building renovation that saves nothing.

Distributed electricity generation is taking over the market. This trend makes economic sense, given that three quarters of U.S. household loads average less than 2.4 kilowatts per hour, and three quarters of commercial customers average less than 10 kilowatts per hour. Eleven percent of U.S. electricity is already generated from renewable sources. Denmark generates one-sixth of its electricity using wind power, with a target of 50 percent by 2030. No land-use problems have resulted, and issues related to the intermittence of supply were resolved long ago. Modern wind turbines occupying only 5 percent of four Montana counties could generate 20 percent of U.S. electricity demand. Energy payback times for these technologies are months to a few years.

Fuel-cell vehicles, in addition to their environmental benefits, can also generate electricity for end-users while parked, allowing the average owner of a Hypercar™, a fuel-cell vehicle designed by Lovins et al., to resell enough electricity to the grid to pay up to half the vehicle's lease cost. A fleet of 110 million fuel-cell cars could produce five to ten times as much electricity as is now generated by all U.S. electric utilities.

Developing countries have great potential for conservation and renewables, as they now use energy only one third as efficiently as developed countries. China, for instance, has cut its coal output by about a third in five years and is shifting rapidly to efficiency, gas, and renewables. Indeed, China recently announced a five-year moratorium on nuclear orders.

The market collapse of nuclear power represents "nonproliferation at a profit." Without nuclear power, all the ingredients needed for nuclear weapons by all known methods would no longer be ordinary items of commerce. They would become harder to obtain because ambiguity and civilian cover would be removed. In addition, the United States must move away from nuclear weapons and toward a new security approach based on conflict avoidance and nonprovocative defense.

Comments by Rhodes and Lovins

Rhodes pointed out that Lovins's numbers on renewables (unlike his own) include hydropower, which is not truly renewable. Lovins asked why Rhodes did not include biomass in his estimates, and Rhodes replied that biomass involves burning materials. Lovins explained that biomass can be grown sustainably. Rhodes noted that manufacturing of renewable energy technology creates emissions; the production of carbon dioxide from manufacturing wind turbines is more than the production of carbon dioxide from the equivalent amount of nuclear power capacity. Lovins rejected Rhodes's estimate as off by a factor of ten. Rhodes pointed out that fuel-cell hydrogen is not an energy source but rather a means of energy transfer, and posited that hydrogen would be made from natural gas. Lovins countered that carbon dioxide could be reinjected at the wellhead, an efficient way to use natural gas in a renewable economy. Rhodes pointed out that thirty-eight nuclear power plants are now under construction, but Lovins dismissed them as old orders. Nuclear power, Lovins said, is not a healthy industry, and it will only get harder to find new orders.

Discussion

Milton Leitenberg of the University of Maryland objected that Rhodes and Lovins's long-term social solutions—abolition of nuclear weapons and conflict avoidance—could not be taken seriously. He also asked Lovins to clarify his point about state utility regulation discouraging efficient end-use. Lovins replied that all states but Oregon reward utilities for selling more electricity, policies that should be changed to decouple profit from the volume of sales. As for conflict resolution, Lovins continued, it is in fact taken seriously in the international relations literature. Rhodes clarified that by "abolition" he did not mean uninventing the bomb, but rather having no assembled nuclear weapons on alert. Nuclear-weapon states cannot retain their enormous nuclear arsenals. Lovins agreed.

Bertram Wolfe, retired from General Electric, pointed out that in a future world of 10 billion people, even with per capita demand only one third of current U.S. per capita demand, we would need three times the current world energy supply. This energy supply will be vital to prevent conflict over energy. Japan entered World War II because it needed energy. If renewables are so cheap, why are we not building them now?

Leventhal asked Wolfe if reprocessing and mixed-oxide fuel could wait until fast breeder reactors were required. Wolfe responded that they could, but fast breeder reactors would be needed in the next thirty to fifty years, and breeder-reactor fuel cycles that do not separate plutonium should be supported.

Damon Moglen of Greenpeace International asked Rhodes if his figures included emissions for the entire nuclear fuel cycle. Rhodes agreed that uranium enrichment uses coal-fired electricity, but nuclear heat could be used to process uranium. Overall, nuclear generates less air pollution than fossil fuels. Moglen rejected the pebble-bed modular reactor design as a desperate attempt to draw business to the South African nuclear industry. Rhodes responded that he did not offer the pebble-bed modular reactor design as the future path forward for nuclear power and that it is not necessarily the answer for developing countries.

Nuclear and Alternative Energy Supply Options for an Environmentally Constrained World: A Long-term Perspective, Robert H. Williams, Princeton University

We need extraordinarily rapid growth of new technologies to increase energy supply yet decrease pollution, simply to keep the damage from pollution constant. Oil and natural gas production will peak between 2025 and 2050. Climate change is the most daunting challenge. Annual emissions of carbon dioxide from global fossil-fuel combustion are projected to increase from about 6 gigatonnes of carbon to nearly 20 gigatonnes by 2100.

Our options are nuclear fission, renewables, and decarbonized fossil fuels utilizing carbon sequestration. Nuclear power faces four serious challenges: costs that are typically higher than those of alternatives; concerns about reactor safety; lack of significant progress in dealing with the disposal of radioactive waste; and the nuclear weapon connection to nuclear power. We would need to deploy 5,000–6,000 gigawatts of nuclear power capacity by 2100 to have any major effect on global warming (compared with 2.7 gigawatts under the business-as-usual projection for 2010 of the Intergovernmental Panel on Climate Change).

Proliferation risks are especially difficult to manage when plutonium is recycled or when uranium fuels are enriched to high levels. In a scenario where uranium shortfalls force a shift to widespread deployment of breeder reactors by 2100, along with reprocessing and plutonium recycling, the amount of plutonium circulating in global commerce would be about 5 million kilograms per year, compared with the less than 10 kilograms of plutonium needed to make a nuclear weapon. It would be very difficult to implement technical fixes to improve proliferation resistance. Large nuclear parks under international control would be required.

Wind power is increasing 30 percent annually worldwide. Wind produced 17 gigawatts of electricity in 2000, only 0.25 percent of the total, but capacity is increasing rapidly. Wind-generated electricity already costs less than 5 cents per kilowatt-hour without subsidization, a cost that could decrease to 3 cents per kilowatt-hour by 2010–2015. Total wind potential is 20,000–50,000 terawatt-hours per year, or one and a half to four times the total amount of electricity now generated annually worldwide. The low-end World Energy Council estimate of 20,000 terawatt-hours per year from wind generation represents a reasonable target for 2100. Compressed air energy storage, at a cost of less than 1 cent per kilowatt-hour, allows wind energy to be used for baseload electricity generation. Wind farms would require only 0.6 percent of the land of the inhabited continents to generate 20,000 terawatt-hours annually.

Solar photovoltaic sales are increasing, and costs are decreasing. Central generation using solar photovoltaics now costs 25–35 cents per kilowatt-hour. Solar photovoltaic capital costs could drop to $3 per peak watt by 2005, and the price of central photovoltaic electricity can reach 4.5–5 cents per kilowatt-hour by 2030. Photovoltaics is already cost-effective for home use with mortgage financing and net metering. At a cost of 10–12 cents per kilowatt-hour, home photovoltaic generation is now competitive for ten million homes at the retail rate.

Coal power with near-zero emissions currently costs less than nuclear power in most regions, and costs will decrease. Zero-emission coal is competitive with the estimated price of electricity generated in an advanced boiling water reactor, even assuming a 30-percent decrease in the costs of nuclear power operation and maintenance.

With carbon sequestration, zero emissions for fuels used directly are possible. At about $1 per gallon of gasoline-equivalent, hydrogen from fossil fuels using carbon sequestration technology is the least costly option, including nuclear, and will become even cheaper with the development of hydrogen-producing membrane reactors.

Alternatives and renewables need to be embraced enthusiastically by the public to succeed. Nuclear power, however, faces problems with public enthusiasm.

A World with, or without, Nuclear Power? Richard L. Garwin, IBM Fellow Emeritus

At least five considerations are important when assessing the potential expansion of nuclear power: proliferation of nuclear weapons; catastrophic accidents; radiation dose to the public from normal operations and the nuclear fuel cycle; global warming; and cost, including capital investment and the fuel cycle.

Proliferation. Uranium enrichment plants, which fuel most of the world's reactors, can be used to enrich uranium to the 90-plus-percent range, an ideal material for nuclear weaponry. Reactors fueled with uranium create plutonium, which has been the material of choice for nuclear weapons. None of the nuclear-weapon states has used reactor-grade plutonium to any great extent for making nuclear weapons. However, it is a myth that nuclear weapons cannot be made using reactor-grade plutonium, a myth that studies by J. Carson Mark, the

National Academy of Sciences, and the U.S. Department of Energy have dispelled.

Even the amount of weapon-usable material generated by the nuclear power industry thus far is almost incomprehensible, enough for tens of thousands of nuclear weapons. The hazard is that a small amount could be stolen, purchased, or diverted to make a few or a few dozen nuclear weapons that could hold even a large country hostage. Mixed-oxide fuel production, for example, entails storage of separated plutonium in canisters that are small and might be easy to carry away. The solution is to account for and to guard the material as discrete items, and not as bulk. Viable geologic waste repositories are also urgently required.

Other possibilities for massive transfer of nominally civilian nuclear materials and capabilities for weapons include failed states. It is not likely that the global community will intervene in a failed state to secure its nuclear installations against sabotage or theft of weapon-usable materials.

Catastrophic accidents. A typical nuclear power reactor core contains long-lived radioisotopes equivalent to 30 megatons of fission. According to 1993 estimates by the UN Special Committee on the Effects of Atomic Radiation, the 1986 Chernobyl accident will ultimately be responsible for some twenty-four thousand cancer deaths—estimates that were curiously deleted from the committee's year 2000 report. Ultimately, the risk of accidents must be weighed against other dangers. Further, reactors such as high-temperature modular gas turbine reactors can help in eliminating the risk of catastrophic accidents.

Radiation dose. In our 1977 Ford-MITRE study, we considered that each reactor-year of operation would involve one to two deaths throughout the nuclear fuel cycle. The nuclear industry claims significant reductions in radioactive emissions since that estimate.

Global warming. Building and operating a nuclear power plant for forty years would contribute less heat to the earth than is provided by a single year's operation of a fossil-fuel plant. Carbon taxes could make coal-generated electricity prohibitively expensive. As Williams shows, however, carbon sequestration can convert inexpensive coal into a benign and flexible fuel for stationary power plants and a hydrogen economy.

Cost. The costs of nuclear power may be far less than assumed by those who believe that uranium shortages will require a move to breeders and plutonium fuel cycles. Indeed, the 4 billion tons of uranium contained in seawater could prove an economically viable resource, adding only about 2 cents per kilowatt-hour to the price of nuclear electricity.

The world could live without nuclear power if coal were consumed utilizing carbon-sequestration technologies. The world, however, could also live with nuclear power if the needed resources are committed to prevent catastrophic accidents and nuclear proliferation.

Discussion

Leventhal asked Garwin whether the nuclear industry risked being shut down completely if there were a catastrophic accident or sabotage incident at a nuclear power plant. Garwin replied that public reaction would depend on how many reactors the nation had. France, for example, would not be able to shut down all its reactors overnight. Transmission lines are also vulnerable.

Thomas Cochran of the Natural Resources Defense Council asked Williams about the difficulties of market penetration for hydrogen and wind. Williams noted that hydrogen would be produced near big cities, using the cheapest feedstock and local distribution. Technologies that he discussed will soon be commercialized for the non-energy production of hydrogen. Lovins agreed that the hydrogen infrastructure is already coming in.

Luncheon Speakers

Representative Edward J. Markey, U.S. Congress

Representative Markey acknowledged Leventhal's and the Nuclear Control Institute's twenty years of contribution to nonproliferation. Markey described legislation that he would soon introduce to provide the Democratic People's Republic of Korea (North Korea) with 2,000 megawatts of coal-fired electricity, instead of the two light-water reactors now scheduled to be supplied under the terms of the 1994 Agreed

Framework. The coal-fired plants would take three to four years to build, in contrast with a decade or more for nuclear plants, and would produce twice as much electricity. The Republic of Korea (South Korea) would pay 80 percent of the cost. The United States can also assist North Korea in modernizing and expanding its electricity-distribution grid. Some have estimated that a modern North Korean electricity grid could cost $1 billion. North Korea's longing for nuclear power plants is like wanting an automobile when you have no roads.

Nuclear Power and Nuclear Weapons Proliferation, Ambassador Robert L. Gallucci, Georgetown University

Should we reprocess spent fuel? Spent fuel remains one of the clear links between nuclear power and the possible spread of nuclear weapons. Proponents cite future uranium shortages, preparation for future fast reactors, and waste management as justification. Opponents cite abundant uranium supplies, debunk the need for fast reactors, and claim that reprocessing complicates waste management.

Should the United States use its leverage to block reprocessing and plutonium fuel? Plans to dispose of weapons plutonium by means of mixed-oxide irradiation forces these issues to be revisited. Plutonium opponents cite the risk of plutonium theft and diversion, while its proponents defend the adequacy of safeguards and the impeccable nonproliferation credentials of plutonium-using allies. This old debate is still not settled.

What are the relative merits of international mechanisms to prevent proliferation (the Nuclear Non-Proliferation Treaty, the International Atomic Energy Agency, and safeguards)? The Iraqi nuclear weapon program was materially advanced by Iraq's treaty status and the agency safeguards. However, there would have been no legal basis for confronting North Korea in 1992 absent its membership in the Nuclear Non-Proliferation Treaty. The limits of the treaty and the International Atomic Energy Agency regime are now well enough understood to make transparent any effort to use the regime to rationalize dangerous technology transfers.

Will any state use its nuclear power program to pursue nuclear weapons? After the first five nuclear weapon states, every proliferant state—North Korea, Pakistan, India, South Africa, Iraq, Iran, Israel, Argentina, Brazil, Taiwan, and South Korea—has pursued nuclear weapons under the cover of nuclear energy, if not a nuclear power program. None, however, planned to divert weapon material directly from a power reactor. Differing assessments of this risk mean the debate continues, particularly over transfers of light-water reactors to North Korea and Iran.

Today's key issues for the future of nuclear power include the following.

1. *Climate change*. The industry is losing its chance to seize this issue by presenting the safest possible nuclear alternatives. Presenting the plutonium option is counterproductive to public acceptance of nuclear power.
2. *Russia*. Access to fissile material has been, and remains, a real barrier to nuclear weapons acquisition by Libya, Iraq, Iran, and perhaps North Korea. It is widely believed that Pakistan, India, Israel, and South Africa all manufactured weapons once they obtained fissile material. Avoiding reprocessing, plutonium, and mixed-oxide fuel is a critical step in the right direction. Russian nuclear exports to Iran and India are also issues of concern.
3. *Countries of particular proliferation concern*. The 1994 Agreed Framework with North Korea seemed preferable to the alternative: war. Non-nuclear alternatives would have been even better, but at the time North Korea would not have agreed to the deal at all if only non-nuclear alternatives were offered. We should now try again to offer North Korea non-nuclear alternatives. They must be cast as a proposal to North Korea, rather than a forced abandonment of the Agreed Framework.
4. *Nuclear terrorism*. No terrorist group on earth is capable of building nuclear weapons, even with assistance. State sponsorship, however, creates a risk if, say, Iraq got hold of fissile material. We know that Iraq has the capability to manufacture a simple fission weapon, lacking only the fissile material, and might not be deterred from transferring nuclear bombs or material to a terrorist group.

In summary, the degree to which nuclear power is connected to nuclear weapons depends on the policy choices of governments.

Discussion

Leventhal noted that the 1986 International Task Force on the Prevention of Nuclear Terrorism, sponsored by the Nuclear Control Institute, commissioned a study by five former nuclear-weapon designers, who concluded that terrorists could indeed build a nuclear weapon. Gallucci conceded that he could not second-guess nuclear-weapon designers on such matters.

Lovins pointed out that there is strong evidence on the market viability of non-nuclear alternatives, and Gallucci concurred.

Wolfe noted that South Africa got five bombs without a nuclear energy program. Gallucci replied that South Africa developed uranium-enrichment technology for the express purpose of developing light-water reactors and entering the global enrichment market. South Africa's nuclear program was indeed publicly portrayed as being directed at civilian nuclear power.

Markey argued that Iran and Iraq do not need nuclear power. By definition, there is a clandestine agenda in those nations. He asked Gallucci why North Korea rejected the fossil-fuel option in 1994, and whether we could renegotiate the Agreed Framework now to provide non-nuclear alternatives to light-water reactors, possibly with a sweetener. Gallucci responded that people differ on whether Iran has a nuclear-weapon program. The extreme inaccuracy of Iran's ballistic missiles makes their nuclear intent clear; those missiles are not meant to deliver conventional weapons. As far as Gallucci could recall, during the 1994 negotiations North Korea said that if it were to give up its gas-graphite reactor program, it wanted the best possible nuclear technology in return, that is, light-water reactors. Gallucci did not think that was because North Korea wanted to use those reactors to build nuclear weapons. He concluded that we need to start any discussions on non-nuclear alternatives with South Korea, rather than North Korea.

Rhodes pointed out that ten years after the Gulf War Iraq has still not been able to get the few kilograms of plutonium needed for a bomb. Garwin added that Iraq had viewed its Osirak reactor (destroyed in a 1981 Israeli air strike) as a source of nuclear weapons, either from its highly enriched uranium fuel or the plutonium it would produce. We cannot say with assurance that Iraq either does or does not have fissile material right now.

Joseph Egan of the Non-Proliferation Trust asked what the United States can credibly say against Russia's providing light-water reactors to Iran, given that the United States is providing them to North Korea. Gallucci replied that during the Agreed Framework negotiations, Viktor Mikhailov, then head of the Russian Ministry of Atomic Energy, said he thought that the West should pay for Russian light-water reactors to be built in North Korea. This idea was rejected. Light-water reactors for Iran would only create proliferation problems; those reactors for North Korea, Gallucci claimed, would help solve proliferation problems.

Panel: Can Nuclear Power Be Made Proliferation-Resistant and Free of Long-Lived Wastes?

Attempts to Reduce the Proliferation Risks of Nuclear Power: Past and Current Initiatives, Marvin Miller, Massachusetts Institute of Technology

The euphoria generated by President Dwight D. Eisenhower's 1954 Atoms for Peace initiative reassured most people that safeguards could effectively minimize the risks of proliferation. The United States and Soviet Union provided nuclear technology and education to other nations, despite the realization that some of the same technologies, materials, and manpower could be applied to making weapons and that safeguards could not prevent such diversion.

The wake-up call on the linkage between the peaceful and military atom was India's test in 1974. Ensuing efforts to minimize the risk that civilian nuclear activities could be used as a cover for a weapon program encompassed both national and international initiatives.

The concern of the Ford and Carter administrations stemmed from the 1974 India test and the prospect of the rapid spread of highly enriched uranium and plutonium, as well as the uranium enrichment and reprocessing technologies that can produce these materials. The Carter administration preferred that any reprocessing or enrichment take place under international or multinational control and sought to implement these views by domestic legislation and international persuasion, with a focus on eliminating the commercial use of plutonium. The only nuclear fuel cycles considered to be proliferation-resistant were those in which neither highly enriched uranium nor plutonium was used in separated form.

The Europeans and Japanese responded that only reprocessing of spent fuel and plutonium recycle and breeding could avert future uranium shortfalls. They claimed that the proliferation risks of plutonium use were exaggerated because it is difficult to make reliable nuclear weapons using reactor-grade plutonium. The weapon-usability of reactor-grade plutonium continues to cause controversy despite the work of knowledgeable individuals such as the late J. Carson Mark, Garwin, and others. The most recent unclassified summary of the U.S. Department of Energy concludes that nuclear weapons could be built using reactor-grade plutonium that would have an assured, reliable yield of one or a few kilotons and a probable yield significantly higher than that. Moreover, there are weapon designs that are "predetonation-proof," that is, they work with any isotopic mixture of plutonium.

It might be possible to make access to reactor-grade plutonium more difficult by modifying reprocessing technology so that plutonium would not be separated or only partially decontaminated. Such schemes were assessed during the Carter administration and found not to offer significant nonproliferation advantages. Proposals to restrict weapon-usable materials and related technologies to international or multinational energy centers were not adopted. The technical, economic, and institutional difficulties involved in setting up and operating such centers remain considerable.

Attempts to increase proliferation resistance continued by means of such designs as the integral fast reactor and thorium-uranium fuel, but their nonproliferation value is debatable.

Failure of advanced uranium-enrichment technologies casts doubt on future cost reductions for nuclear power. Seawater uranium could increase nuclear electricity costs by 15 percent, but costs need to be reduced by 30 percent if nuclear is to compete with gas turbines.

Proliferation-resistant technologies may require international or multinational energy centers to be politically acceptable. Although such arrangements would be extremely difficult to implement and could create tension between nuclear and non-nuclear states within the Nuclear Non-Proliferation Treaty regime, international control is probably the only way, short of the abandonment of nuclear power, to break the power/proliferation linkage.

Technical Opportunities for Increasing Proliferation Resistance of Nuclear Power Systems (TOPS) Task Force, James A. Hassberger, Lawrence Livermore National Laboratory

One of the premises of the Department of Energy's Technical Opportunities for Increasing Proliferation Resistance of Nuclear Power Systems (TOPS) task force was the recognition that continued reliance on nuclear power was likely in the near term in the United States and throughout the developed world, and that expansion of nuclear power globally was likely.

The task force reached consensus on three points. First, there is both the potential and the need to perform proliferation-resistance research and development. Since the last large-scale reviews of proliferation-resistance technologies in the 1970s and 1980s, there have been significant advances in numerous technologies that might be brought to bear on this problem. Rather than singling out one system, TOPS noted several promising possibilities. Second, proliferation-resistance is only one of several challenges facing nuclear power and must be considered in the broader contexts of nuclear power economics, safety, and waste disposal.

Third, we must explore other opportunities for reducing the risk of proliferation worldwide. We have to look at continuing to reduce the risk of theft of weapon-usable materials, particularly in places like Russia. We must continue to strengthen international safeguards. The bulk of international safeguards funding is spent safeguarding plants in weapon countries and developed states, resources that could be more efficaciously spent safeguarding facilities elsewhere in the world.

Various approaches to proliferation resistance include strengthening institutional safeguards, making weapon-usable materials inaccessible, reducing the attractiveness of materials for weapon use, and limiting the spread of weapon-usable knowledge and skills.

There is currently no systematic approach to evaluating either the risk of proliferation or the efficacy of options to reduce it. TOPS concluded that it is important to develop mechanisms to conduct these evaluations, as well as to examine various approaches to improving the effectiveness and efficiency of safeguards. Making weapon-usable material inaccessible and reducing the quality or the attractiveness of such materials both contribute to proliferation resistance. The question then becomes, how good is good enough?

Finally, we must evaluate the range of technical options and fuel cycles that can start meeting some of these objectives. Such efforts require broad international participation and must be done in concert with meeting the other challenges facing nuclear power.

TOPS developed a "barriers approach" to proliferation resistance, taken broadly from work of the National Academy of Sciences. Technologies should be developed to strengthen both extrinsic barriers (institutions, safeguards, and physical security) and intrinsic barriers (features inherent to the fuel cycle and materials and technologies).

Fissile materials are the key link between commercial nuclear power and weapons. TOPS noted that, historically, civilian nuclear power has not been the path of choice to nuclear weapons, although it has perhaps been used as cover for covert activities. The isotopic and chemical qualities of materials used in the commercial nuclear fuel cycle do not make them ideal for use in nuclear weapons and are barriers to proliferation. However, the fact that someone can probably make a nuclear weapon out of reactor-grade plutonium limits that intrinsic barrier, and so we carefully apply extrinsic barriers to help compensate.

Extrinsic barriers depend heavily on implementation details such as the existence of treaty regimes. One of the most important things is to improve the extent and the effectiveness of extrinsic barriers. The near-term opportunities focus on improving extrinsic barriers, institutional measures, safeguards, and monitoring.

The Limits of the Technical Fixes, Edwin S. Lyman, Nuclear Control Institute

The notion of proliferation resistance is practically gospel in some quarters, including the U.S. Department of Energy and Russia's Ministry of Atomic Energy. If the notion of proliferation resistance is intended merely to put a nonproliferation seal of approval on plutonium recycling, it will be extremely counterproductive. To have a truly proliferation-resistant closed fuel cycle, the risk should be no greater than that of the once-through cycle. Achieving this "spent fuel standard" will not be simple and would likely raise the costs and risks of the technology to unacceptable levels. However, anything less will at best provide a marginal reduction in risk, and at worst provide false confidence that will greatly increase the dangers of proliferation in the long run.

The most important barrier against theft of spent fuel is its "self-protecting" radiation field. However, this barrier inevitably decreases with time, eventually no longer providing adequate self-protection. One way to mitigate the resulting risk is to emplace spent fuel irretrievably in a repository, providing a "geologic" barrier.

Some argue that the risk of mining plutonium from repositories for use in weapons is unacceptably high. However, a study by Lyman and Harold Feiveson showed that the attractiveness of such schemes would be relatively low and the mining would be easy to detect. Nevertheless, conventional nuclear power plant operation is creating a massive plutonium inventory that will pose risks far into the future.

It is unlikely that any once-through system could effectively reduce weapon-usable material to a level below concern. Further, the near-term proliferation risks of the necessary additional reprocessing would be significant. Safeguards difficulties increase dramatically as the focus shifts from item-counting to material accountancy at bulk-handling facilities, in which statistical errors and measurement biases can create large uncertainties that can serve as a cover for the diversion of materials. The dirtier the material, the less accurate the assays become. Any system will have diversion pathways that can defeat containment and surveillance; thus, there is no substitute for material accountancy.

Proposed concepts to decrease the plutonium content of spent fuel promise only about a fivefold reduction, far short of the hundredfold reduction required to have a meaningful impact. The radiation barriers provided by residual fission products in most proposed systems are oversold. Proliferation-resistant technologies also raise many safety issues. The goal of modifying systems to *increase* the radiation hazards as a deterrent to theft conflicts with the goal of protecting workers, the public, and the environment.

A reduction in the risk of proliferation can be most easily achieved through the centralization of nuclear facilities and materials. Broad deployment of small gas-turbine nuclear plants would require credible means of reducing the associated risks of proliferation without overwhelming the safeguards system.

Perhaps the greatest obstacle to the expansion of nuclear power plants, especially to regions of political instability, is the threat of radiological sabotage. An armed assault on a nuclear plant can result in core melt, containment failure, and a massive, Chernobyl-like release of radioactive materials. Radiological sabotage could conceivably fulfill the same goals for terrorists as the acquisition and use of a crude nuclear weapon. The risk of radiological sabotage is of particular concern in politically unstable regions in the developing world, those targeted as the most likely customers for small, modular nuclear plants.

The proposal to eliminate nuclear waste by means of accelerator transmutation of waste has powerful congressional allies. However, the accelerator transmutation of waste road map of the Department of Energy estimates that transmuting the entire U.S. spent fuel inventory would take 118 years, cost some $279 billion, and require enormous investment and oversight by the federal government.

Discussion

Garwin asked if modular reactors could be sited underground. Lyman replied that this approach is not being examined for the pebble-bed modular reactor and also would be extremely costly.

Cochran asked if there is any value to federal research and development on accelerator transmutation of waste and whether there is any report that accelerator transmutation of waste would reduce the health effects of nuclear waste. Miller responded that accelerator transmutation of waste does not buy very much in that regard. Hassberger said that we need to consider how to balance the long-term risks of nuclear waste left untreated versus the local risks of accelerator transmutation of waste processing facilities, and how society should value future lives versus present lives. Lovins commented that such a calculus would be immoral.

Panel: The Role of Nuclear Power in the Acquisition of Nuclear Weapons

Overview of Nuclear Power and Nuclear Weapons, Zachary S. Davis, Lawrence Livermore National Laboratory

The nuclear nonproliferation regime is a combination of domestic laws, international institutions, technical arrangements, and bilateral agreements—all held together by skillful diplomacy and a little smoke and mirrors. This panel evaluates the effectiveness of technical and legal barriers established to maintain separation between civil and military applications of nuclear technology.

George Perkovich presents case studies of India, Pakistan, and Iran. There are cracks developing in the nonproliferation regime. For example, Russia is disregarding the Nuclear Suppliers Group's policy of requiring full-scope safeguards as a condition of supply so that it can sell reactors to India. What are the long-term consequences of such violations? How will the breakdown of Nuclear Suppliers Group standards affect proliferation elsewhere?

In the case of India, we can look backward to see how civil nuclear technology contributed to its nuclear weapons program. Iran requires us to look forward. Will Iran follow India's example and use its civil nuclear infrastructure to develop nuclear weapons? Iran (unlike India) is a party to the Nuclear Non-Proliferation Treaty, but there are questions about the long-term health of the nonproliferation regime and of the treaty itself.

Lawrence Scheinman examines whether the nuclear nonproliferation regime is in trouble and assesses the future viability of legal, multilateral, and technical barriers. Reforms have been instituted to revitalize the International Atomic Energy Agency and make the safeguards system more robust, intrusive, reliable, and credible. Questions persist, however, about the "93 + 2" Additional Protocol to State Safeguards Agreements intended to strengthen the safeguards system. Given its meager budget, how can the International Atomic Energy Agency safeguards be expected to expand to include more fissile material, more facilities, and a wider range of activities? Stresses are becoming increasingly evident at the Nuclear Non-Proliferation Treaty review conferences, so we must also ask whether the treaty regime is in trouble.

The panel concludes with a response by Leventhal, who presents his assessment of the performance of U.S. nonproliferation laws and policy and discusses the challenges ahead for the regime.

Nuclear Power and Weapons in India, Pakistan, and Iran, George Perkovich, W. Alton Jones Foundation

Perkovich develops four basic themes regarding how nuclear power programs provide cover and comfort for nuclear-weapon programs: diversion of materials; training, technology, and procurement; general cover; and creation of a bureaucracy or establishment that can affect a nation's nuclear-weapons decision making. All four themes apply to India. Some also apply to Pakistan and Iran. In the case of India, much of the nonproliferation damage occurred before the Nuclear Non-Proliferation Treaty was even negotiated. Further, India is not a treaty member state. India did violate restrictions in its nuclear cooperation agreements with Canada and the United States.

Diversion of materials. In 1956, India sought to weaken controls in the International Atomic Energy Agency statute then under negotiation. At the time, Homi Baba, the driving force in India's early nuclear program, noted that a country could develop a "parallel program" not subject to safeguards, but still draw on the expertise and resources of the public nuclear program to produce nuclear weapons. India violated nuclear cooperation agreements with the United States and Canada when it procured the plutonium for its 1974 nuclear test from reprocessing fuel from its CIRUS reactor. Blame must be shared, however: Government and nuclear industry leaders in the United States,

Canada, France, and elsewhere were culpable "enablers" by promoting nuclear power in India.

Training, technology, and procurement. The United States trained more than one thousand Indian nuclear scientists and engineers prior to 1972. Some worked in the Indian weapon program. Indian scientists learned about neutron-initiator research, useful in the development of nuclear weapons, while working in French labs, and Pakistani scientists learned about it working in China. All the key figures in India's nuclear weapon program have also worked simultaneously on civil projects; they were "dual-use people." Munir Kahn once told the author that Pakistan pursued reprocessing and fuel cycle technology for Pakistan in the 1970s so that France would build the infrastructure and train hundreds of engineers, allowing Pakistan to construct and operate separate, parallel, unsafeguarded plants. "If I don't get the cooperation, I can't train the people to run a weapons program," Kahn said. This experience provides a lesson for the risks in Iran today.

General cover for a nuclear weapon program. The larger a nation's civilian nuclear program, the easier it is to hide weapon work. Legitimate civilian activities shield harder-to-detect military activities. A power reactor is the least of the problems in Iran. An agreement with Iran should be negotiated that would bar reprocessing, enrichment, or deuterium facilities.

Bureaucratic influence. Once nuclear establishments are formed in states, the "wizards" responsible for building these expensive, complicated facilities achieve mystique and become national heroes. This phenomenon has been most apparent in India and Pakistan. Other examples of nuclear bureaucratic influence include the power of the Ministry of Atomic Energy in Russian domestic politics and the role of U.S. nuclear labs in determining national policy on the comprehensive test ban issue. Nations can be moved along the path to weapon work by the bureaucratic sway of their civil nuclear sector. Iran's nuclear establishment does not yet have the mystique or power of the nuclear establishments in India and Pakistan.

We must also account for the danger of "reverse military conversion." Nuclear programs could shift from civilian to military applications as options for additional civilian nuclear development are foreclosed. This could be a risk in such nations as Iran, Japan, and

South Korea. Given the enormous costs sunk into training and facilities, politicians might be tempted to convert these assets to military applications rather than let them languish.

The Nonproliferation Regime and Fissile Materials, Lawrence Scheinman, Monterey Institute for International Studies

The nonproliferation regime is being challenged, but we must ask why states joined the regime in the first place. If these reasons are still valid for a given state, that state is likely to remain in the regime. Articles 4 and 6 of the Nuclear Non-Proliferation Treaty are both implicit bargains. The treaty regime also provides security, stability, and predictability for non-nuclear-weapon state signatories.

The most important roadblock to proliferation is access to fissile material. Any kind of plutonium should be regarded as a proliferation risk and be dealt with accordingly. A radical alternative would be to forswear using nuclear energy, but that is not a viable option for several reasons. Neither institutional nor technical strategies alone are sufficient. Both strategies require unwavering political commitment; greater prioritization for nonproliferation; leadership by key nations to strengthen the regime; and determined collective response in cases of noncompliance.

Even if we eliminated nuclear power, the risks of proliferation would not end. States can and have pursued nuclear weapons without a peaceful nuclear program. Nor does global nuclear disarmament and disposal of all attendant fissile materials seem likely.

Experience shows that nonproliferation regimes do matter. They erect legal barriers; embody nonproliferation norms; raise the stakes of acquiring nuclear weapons; build confidence; and provide a framework for export control, verification, and collective responses. While not alone sufficient to prevent the spread of nuclear weapons, they are a necessary element. Delegitimization of nuclear weapons, arms control, and an effective collective security system also relate to regime credibility.

Iraq's clandestine nuclear weapon program underscored shortcomings in the safeguards system of the International Atomic Energy Agency. Improvements have been made to agency safeguards since then, including the 93 + 2 Additional Protocol, substantially raising the bar that a state subject to comprehensive safeguards would have to scale to acquire nuclear weapons. However, more states must bring the Additional Protocol into force, and the safeguards system must be adequately funded. It would be a pyrrhic victory to have a strengthened safeguards system without adequate funding, yet that is the situation the International Atomic Energy Agency faces today.

We need to determine best approaches to minimize separated plutonium or highly enriched uranium, including technical options that eliminate the separation of plutonium from spent fuel; control the separated plutonium effectively as we reduce stockpiles; pace any further reprocessing to avoid plutonium surpluses; and secure spent fuel pending disposition. These approaches suggest building out from the existing regime structures to fashion new institutional arrangements.

Non-nuclear-weapon states with substantial stocks of separated plutonium, particularly Japan, have become increasingly sensitive to how these stocks are perceived and whether arrangements might be crafted to enhance security and ameliorate concerns. This concern offers an opportunity to revisit international plutonium and spent-fuel storage proposals.

The nonproliferation regime is a dynamic system amenable to growth and innovation. Together with identification of acceptable ways to permit access to nuclear technology for peaceful purposes without incurring proliferation risks attendant to current fuel cycle choices, innovations may reduce fissile-material risks. Other important nonproliferation goals include negotiation of a fissile material cutoff treaty and expanded access by the International Atomic Energy Agency to intelligence information.

Closing Thoughts on Nonproliferation: The Need for Rigor, Paul L. Leventhal, Nuclear Control Institute

Adrian Fisher, first general counsel of the Arms Control and Disarmament Agency and the chief U.S. negotiator of the Nuclear Non-Proliferation Treaty, once said that the treaty does not obligate us to do anything foolish. Nothing in the treaty requires us to go against our supreme national security interests, for the sake of adherence to the letter of the treaty, by exporting nuclear materials and technology to nations suspected of wanting the bomb. The treaty can be interpreted in a number of ways to fulfill solemn national security interests.

There is a "dynamic tension" between Articles I and II and Article IV of the Nuclear Non-Proliferation Treaty. A transfer of nuclear technology authorized by Article IV cannot happen if it is not "in conformity with" Article I and II obligations not to do anything to transfer nuclear weapon capability to non-nuclear-weapon states. A 1995 legal analysis by Nuclear Control Institute counsel Eldon Greenberg made the point that if there were no economic necessity justifying the use of plutonium as fuel, then the treaty, as it presently existed, and without any need to amend it, could be interpreted to say that separation and utilization of plutonium run afoul of Article I and II prohibitions.

We must ask if the continued utilization of plutonium and highly enriched uranium makes any sense at all in terms of the future of the nuclear industry and world security. The Nuclear Non-Proliferation Treaty regime today lacks a sense of rigor, and the regime is being applied for the purpose of legimitating commerce in weapon-usable plutonium.

The approach today is to find ways to allow plutonium to be used, rather than develop a consensus that plutonium is too dangerous to use and build a treaty regime to prevent that danger. The regime does not deal with plutonium for what it is—an atom-bomb material too dangerous to be used in commerce. The nuclear industry and bureaucracy control deliberations on the Nuclear Non-Proliferation Treaty, so there is also a pronuclear industry bias that pervades nonproliferation meetings and makes dealing with the dangers of plutonium all but impossible.

Why has highly enriched uranium largely been phased out, but not plutonium? A near-consensus to get rid of highly enriched uranium has been possible because there is no industrial bias to retain it.

The Nuclear Control Institute proposed in 1994 that nuclear-weapon states and non-nuclear-weapon states could converge on the use of immobilization to dispose of surplus plutonium. A decision by the United States and Russia to immobilize their surplus military plutonium in highly radioactive waste for eventual geological disposal could be expanded into a multilateral regime, in effect creating a "magnet" to draw in separated civil plutonium as well. Unfortunately, the Russian Ministry of Atomic Energy and the U.S. Department of Energy both wanted to utilize plutonium disposition to revitalize their plutonium industries by turning surplus weapon plutonium into mixed-oxide fuel for use in commercial power reactors. This approach sets precisely the wrong example for nonweapon states.

One can be against plutonium without being antinuclear. Indeed, by avoiding public backlash against plutonium hazards, opposing plutonium might even prove to be pronuclear. Control of nonproliferation policy must be taken away from plutonium zealots. Market forces might yet do the trick. We need some sort of catalyst, hopefully short of catastrophe, to bring sense to the nonproliferation regime.

Discussion

Cochran noted that the nonproliferation regime failed in India, Iraq, and North Korea. He asked Davis whether there is any evidence of Russia's providing training, advanced fuel cycle technology, or even nuclear weapon technology to Iran. Davis demurred, saying he could not go beyond what has been reported in the press. There could still be a positive outcome in which Iran remains in the Nuclear Non-Proliferation Treaty and safeguards regime. Perkovich said there will be more nonproliferation pressure within Iran, and hence more leverage available to the United States, if Iran remains publicly committed to the treaty. Scheinman added that no solution is possible to the Iranian nuclear problem unless Iraq's nuclear bomb program is addressed, and there have been no inspections in Iraq for over two years.

Cochran asked Perkovich if he was claiming the United States should "cover up" any Russian-Iranian violations of the Nuclear Non-Proliferation Treaty it discovered. Perkovich disagreed; his point, he said, was that quiet diplomacy could be more effective than igniting an international controversy over Iran's nuclear ambitions. Leventhal noted that North Korea, after joining the treaty in 1985, failed for five years to provide information on its nuclear facilities to the International Atomic Energy Agency, yet Russia asked the United States not to make an issue of it.

Leitenberg asked Scheinman about the likely effect on the regime of the UN Security Council's failure to back up inspections in Iraq by the UN Special Commission on Iraq. Scheinman said we will have to wait and see whether Bush administration policy will offer credible support for nonproliferation or focus its efforts primarily on "counter-proliferation." Feiveson of Princeton University asked what would happen if a nation withdrew from the Nuclear Non-Proliferation Treaty rather than violating it. Scheinman replied that the treaty provides for a ninety-day notice, time to cool off, and an explanation of the decision to withdraw. A state would need to make a case justifying its withdrawal. The UN Security Council could still stand firm and reject the explanation, citing withdrawal from the treaty as a threat to international security, or offer positive security guarantees to bring the party back into the regime. Perkovich added that Iran could present a compelling case for its national security withdrawal from the treaty based on the documented nuclear threat from Iraq, putting the five permanent members of the Security Council in difficult positions.

Roundtable Discussion: Three Closing Views

An Industrialist's View, Bertram Wolfe, General Electric Corporation, Retired

After 1954, when we started building nuclear plants in this country, there was a tremendous need for energy here. We were rapidly expanding energy in the late sixties and early seventies, doubling energy use every ten years. There were tremendous orders for both coal and nuclear plants. We were selling thirty to forty nuclear plants a year, and we expected to have over a thousand nuclear plants operating by the end of the century.

Since 1973, an abundance of generating capacity has allowed U.S. environmentalists the luxury of opposing nuclear power. The United States has not needed lots of new capacity until recently, but now we do. New energy sources are required; three times more energy will be needed to power a world of 10 billion people.

Reprocessing and mixed-oxide fuel are not needed to fuel light-water reactors, but we will run out of economic uranium resources in the next fifty years. We should address the issue of nuclear waste on an international basis. Later, we should develop international reprocessing centers for plutonium-fuel use in safe fast reactors. To wait until the last minute to develop these systems would be devastating.

An Arms Controller's View, Harold A. Feiveson, Princeton University

It was once imagined that a once-through nuclear power system would be proliferation-resistant. What we worried about was separation of plutonium and the onset of breeder reactors. But this large-scale future for nuclear power that is now being seriously discussed makes me most uneasy. As Williams noted, the nuclear power system would have to grow to ten to twenty times its present size if it is to make any substantial dent in the greenhouse problem. As to nuclear waste, there could be technical solutions, but a nuclear power system of that scope would produce roughly one Yucca Mountain of waste per year.

Even if nuclear power spreads throughout the world, it does not necessarily mean that it will become less safe. As to the question of proliferation resistance, a vastly expanded civil nuclear industry would give us a tremendous legacy of fissile materials that must be safeguarded forever.

Garwin and others have suggested once-through fuel cycles drawing upon the enormous uranium resource dissolved in seawater. However, to fuel a vastly expanded civil nuclear power section in this way would require widespread proliferation of uranium enrichment plants. Something like six hundred and fifty 1,000-ton uranium enrichment plants around the world would need to be working each year. Each plant, if it started with natural uranium, would make three hundred bombs' worth of uranium. If it started with 8-percent-enriched uranium, it could make maybe six times that amount, about eighteen hundred bombs per year.

In that world, you would have a tremendous amount of uranium coming out of seawater and tremendous incentive to innovate to make uranium enrichment cheaper, quicker, and faster. It might not be impossible to safeguard that kind of system, but it does give one pause.

In the long run, if you are going to have nuclear power of this magnitude, you probably have to base it—and maybe even this would not work—on very centralized nuclear parks under international control. This means that countries would have to give up a great deal of sovereignty over their nuclear energy systems.

A Historian's View, William Lanouette, U.S. General Accounting Office

What will it take now to elevate the issue of nuclear proliferation in the context of nuclear power so that the public—and certainly the responsible policy people—really give it some concern? Will it take yet another state going nuclear, a publicized theft of nuclear material, or some scandal involving nuclear secrets? Cases such as Mordecai Vanunu and Wen Ho Lee receive extensive media coverage, but how do they really affect our consciousness about the nuclear weapon threat?

The end of the cold war should be a time to focus on nuclear power and nuclear weapons. Truly original thinking is required, such as that of Leo Szilard, who once disagreed with the cofounders of the Pugwash Movement for nuclear arms control, who favored a ban on nuclear tests. Szilard said, "If you want to stop this whole thing, you've got to test them, test them *all*." Truly original, even irreverent, thinking may get us through the next nuclear era.

Discussion

Williams noted that the world would need to add 100 gigawatts of nuclear capacity a year for a hundred years to make even a dent in global warming, which is totally impossible without widespread public enthusiasm for nuclear power. There is currently no credible path forward for nuclear power deployment on this scale. We should continue to analyze the proliferation risks of utilizing nuclear power to address climate change, particularly the risks of disseminating uranium-enrichment technology. It is also important to draw the more general energy community, not just nuclear experts, into this discussion.

Linda Gunter of the Safe Energy Communication Council added that, contrary to Wolfe's remarks, it is logical for environmentalists to oppose nuclear fission because its damage is obvious. Renewable energy provides more sensible alternatives.

Garwin commented that he favors Wolfe's international repository proposal and would support plutonium recovery in the future if we run out of other economical energy resources. Safeguarding enrichment of uranium from seawater would not pose a major problem. Garwin characterized the TOPS-report participants as "at the trough," in search of funding for their preferred nuclear technologies. TOPS ignored the potential to extract uranium from seawater, and Wolfe's concerns that uranium will run out fail to account for this resource. Seawater uranium-fueled light-water reactors would be cheaper than reactors fueled with recycled plutonium, according to Garwin; Wolfe disagreed.

PART I

How Essential Is Nuclear Power?

3

Nuclear Power and Nuclear Weapons Proliferation

Ambassador Robert L. Gallucci

Some people believe there is nothing new to say about the connection between nuclear energy and nuclear weapons. That assertion seems odd in light of all that has happened over the last fifty years, and particularly over the last twenty-five years. During the mid-seventies any number of debates raged among the experts. Is it possible that the world has learned nothing from a quarter century of experience? Is it possible that the evidence has changed no minds?

It is useful to look briefly at four old debates that helped define today's appreciation of the links between nuclear power and nuclear proliferation and to offer some observations about their status. It is also important to look at four new issues that ought to define the connection now.

Four Old Debates

The first of the old debates involved the *virtues of reprocessing spent nuclear fuel.* Proponents argued that reprocessing made sense for some, if not all, countries with nuclear power programs as a way to:

1. recover plutonium for recycling in thermal reactors, conserve uranium (a scarce resource), avoid wasting the energy value of plutonium, and save on enrichment work;
2. earn money by offering reprocessing services to other countries;
3. learn about and be ready to enter the coming world of fast reactors;
4. help manage radioactive waste; and
5. increase the contribution that nuclear energy could make to a country's energy independence.

Opponents argued that separating plutonium from spent fuel and allowing the technology to spread increased the risk that it would be used to manufacture nuclear weapons. Incurring that risk was unnecessary because:

1. there was no need to recover plutonium since there was plenty of uranium, and the cost of recycling far outweighed the savings;
2. reprocessing services did not yield a net profit;
3. fast reactors were not becoming available;
4. reprocessing made radioactive waste management more rather than less difficult; and
5. energy independence was a myth for most countries and not significantly affected by nuclear power in any case.

It appears that this old debate is still raging, notwithstanding the years of experience that should have resolved most of the points in dispute. Moreover, most of the proponents and opponents are still wearing the same uniforms they wore in 1975. Some quarters in Europe and Japan apparently continue to be interested in plutonium fuels, and Russia is certainly not enthusiastic about treating as waste the plutonium that is contained in spent fuel or that has already been separated. The emphasis on different elements of the old arguments may have changed, but the bottom line is that there is still no worldwide consensus on the issue. It must be concluded that reprocessing and the use of plutonium fuels remain one of the clear links between nuclear power programs and the possible spread of nuclear weapons.

The second old debate was a derivative of the first but with a focus on *the policy of the United States*: Should this country, as a matter of policy, use whatever legal and political leverage it has to block plutonium separation and its use as fuel? Proponents argued that:

1. diversion of plutonium by the government of any non-nuclear-weapon state and by terrorists and criminals was a concern;
2. diversion by terrorists and criminals was a concern even in a nuclear-weapon state;
3. the separation and use of plutonium fuels by any state provided a rationale and degree of legitimacy for any other state to do the same; and
4. no amount of physical security would ever provide sufficient confidence that the very small amounts of plutonium needed to make a nuclear weapon would not be diverted from a program involving large flows and transport of plutonium.

Opponents argued that:

1. distinctions can and should be made between those non-nuclear-weapon states that have no intention of diverting plutonium and manufacturing nuclear weapons, and those that can be expected to secure it through unauthorized access;
2. the choices of nuclear-weapon states relating to civil use of plutonium posed no additional concern;
3. by attempting to impose its views of how advanced states should meet their energy needs, the United States lost the much needed cooperation of those states and damaged its relations with them, so that the effort to do so was counterproductive; and
4. there was no reason to believe that plutonium fuels could not be used without theft.

It appears that not many minds have changed on this issue, even though in most cases policy has settled the matter.

One very interesting current case involves Russia and the agreement that some separated plutonium be used as light-water reactor fuel there and in the United States in the interest of disposing of it and encouraging the disposal of additional separated plutonium by treatment as waste. This case seemed to trouble many who had made up their minds on the subject but found themselves unwilling to allow the best (avoiding any use of plutonium as fuel) to become the enemy of the good (burning up plutonium in fuel to get rid of it), so to speak. For the United States, how to deal with its friends' decisions about plutonium use is still an issue, and plutonium remains a connection between nuclear energy and nuclear weapon concerns.

The third old debate was over the *merits of the international mechanisms established to prevent nuclear proliferation*, particularly the Nuclear Non-Proliferation Treaty and the International Atomic Energy Agency safeguards. Proponents argued that:

1. the Nuclear Non-Proliferation Treaty was a nearly universal agreement that created the essential norm for preventing proliferation; and
2. the safeguards system it mandated was a critical deterrent to diversion.

Opponents argued that:

1. the Nuclear Non-Proliferation Treaty was a weak reed because Israel, India, and Pakistan remained outside the regime and Libya, Iraq, Iran, and the Democratic People's Republic of Korea (North Korea) were in it;
2. International Atomic Energy Agency safeguards could never adequately ensure against diversion from reprocessing or enrichment plants and consequently only served to mislead the international community about the risks associated with these facilities; and
3. together with International Atomic Energy Agency safeguards, the Nuclear Non-Proliferation Treaty by specific provision created a rationale for the transfer of nuclear technology to states that were known proliferation risks and that would not otherwise have been candidates for nuclear cooperation.

On the issue of proliferation, it seems that the virtues and limits of the Nuclear Non-Proliferation Treaty and International Atomic Energy Agency are now better understood. It is difficult to argue that both Iraq's status as a party to the treaty and its willingness to accept the agency's safeguards have not materially advanced the Iraqi nuclear weapon program. This has become clear in the aftermath of the Gulf War as a result of the information that has emerged about Iraq's nuclear weapon calculations in the late seventies and the extraordinary program it pursued throughout the eighties. At the same time, were it not for North Korea's treaty obligations and the agency's inspections conducted pursuant to them, the legal basis for confronting that country at the International Atomic Energy Agency and then the UN Security Council would not have been available. The implication is that the treaty and agency are critical elements in international efforts to manage the nuclear energy–nuclear weapon connection, and that their limits are now well enough understood to make completely transparent any effort to use them to rationalize dangerous technology transfers.

The fourth old debate was over *the likelihood that any state would actually use its nuclear power program to pursue a nuclear weapon program.* Those who said it would argued that:

1. nuclear energy legitimized the pursuit of all the technologies necessary for the acquisition of the fissile material needed for nuclear weapons; and
2. from a proliferation perspective, there was no safe nuclear technology.

Those who said it would not argued that:

1. states would find it substantially easier to build dedicated facilities for a nuclear weapon program than to divert material from a nuclear power program; and
2. it was important to distinguish between fuel cycles and technologies that are more or less proliferation-resistant.

This argument, even as captured above, was always and continues to be sloppy, notwithstanding the experience of the last few decades. Without exception, the non-nuclear-weapon states that seriously explored and pursued the nuclear weapon option—North Korea, Pakistan, India, South Africa, Iraq, Iran, Israel, Argentina, Brazil, Taiwan, and the Republic of Korea (South Korea)—did so under the cover of a nuclear energy, if not a nuclear power, program. However, none of those countries had a plan to divert nuclear material from a nuclear power reactor, which many of them had. Even though the technical issues here are not much in debate, the differing assessments of the risks associated with the transfer of power reactors suggest that the debate continues, as in the cases of North Korea and Iran. The evidence of the connection between nuclear power and nuclear weapons cannot be denied, although the nature of that connection is clearly open to debate.

Four New Issues

Four new issues ought to help shape the discussion of the connections between nuclear energy and nuclear proliferation today. The first is the *concern about climate change*. Agreement is now nearly universal that potentially catastrophic consequences may result from a failure to change current trends in the use of fuels that produce carbon dioxide to meet worldwide energy needs. For this reason, and perhaps for some others, it is possible that governments around the world will see new benefits to including, or increasing the role of, nuclear power in their energy mix to meet national needs. A window of opportunity of sorts may be opening for nuclear energy that, at least in some countries, has been closed for some time. There will be obstacles as always, including capital costs, but there is still an opportunity. Some studies suggest that while nuclear power will not be a panacea (not that anything ever is), it can, in conjunction with technologies that limit emissions and with natural gas and renewables, play a significant role in meeting targets for reduced carbon production, should it be possible to set them. The interesting issue here is, however, the extent to which the public in the United States and elsewhere will prove willing to accept the risks it believes are associated with nuclear power in order to reduce the risks of global warming. The public has special fears about nuclear power. It worries about the immediate and long-term effects of a catastrophic nuclear reactor accident, about poorly stored radioactive waste that slowly poisons water supplies, and about terrorists stealing and spreading nuclear material or using it to make nuclear weapons.

The issue, then, is why governments and the nuclear industry do not together seize this opportunity to offer the public a nuclear power option that is as safe as they can make it? Why do they not emphasize how far the industry has come over the last half century in being able to build and operate plants safely? Why is there not a concerted effort to explain the spent fuel and radioactive waste problem as the genuine political issue it is, rather than as a technical challenge comparable to national missile defense? Why do not industry and the government simplify the real and perceived risks associated with nuclear power by forgoing the use of plutonium in the future and the treatment of spent fuel as waste? The alternative leads people who would not otherwise oppose nuclear energy to argue that a future of many nuclear reactors, where plutonium is preserved in stored spent fuel or separated form, is not one to be embraced, even in the interest of limiting climate change. The decisions of governments regarding the preservation of the plutonium option may either sharply define or substantially eliminate the connection between nuclear power and nuclear weapons.

The second current issue is *Russia and the policies the Russian government adopts at home and abroad.* It is not necessary to repeat the concerns over the security of the fissile material in the former Soviet Union, and particularly the quantities involved in Russia. There are many reasons why the problem of nuclear weapon proliferation around the world is now limited to a relatively few countries, rather than to tens of countries, but none more important than the success to date in limiting the availability of reprocessing and enrichment technology and thus the plutonium and highly enriched uranium those facilities would yield. Little about fission weapons is a secret, and there are fewer secrets about thermonuclear designs than there once were. Fissile material has, however, remained an obstacle to the ambitions of some countries, such as Libya, Iraq, Iran, and perhaps North Korea. It is widely believed that Pakistan, India, Israel, and South Africa all manufactured nuclear weapons once they obtained fissile material. The shape of the threat could change rapidly if the

Russian government does not control its fissile material adequately. If, for example, by theft, criminal activity, or whatever means a few "baseballs" of plutonium find their way across those incredibly long borders, the dimensions of the problem become different very quickly. What is more, were that loss of plutonium to happen, no one should be confident that the world would know which country or group had received it. Indeed, the confidence with which it is asserted that a state or a terrorist organization does not have nuclear weapons ought to be adjusted to fit this new reality.

What is the relevance of all this to nuclear energy? The policies the Russian government adopts can limit or exacerbate these risks. Fully cooperating with concerned governments in programs that improve their ability to account for and secure fissile material are critical. Just as critical over the long term is the need for the Russian government to reconsider its apparent commitment to preserve the option to use plutonium fuels. A decision to forgo further plutonium separation for any purpose and to dispose of separated plutonium in excess of weapon needs as quickly and safely as possible—including burning it in once-through mixed oxide-fueled thermal reactors as well as treating it as waste—would be critical steps in the right direction.

Beyond this issue is the question of Russia's export policies, particularly with respect to Iran and India. In the case of Iran, Russia is well-acquainted with that country's medium and longer range ballistic missile programs, as well as with its interest in fuel cycle facilities beyond what is necessary to operate a light-water reactor. The reasons that Russia should exercise the greatest degree of restraint in its nuclear exports are clear.

In the case of India, the stakes are different but still quite high. The norm of requiring full-scope safeguards as a condition of supply to non-nuclear-weapon states, as the Nuclear Non-Proliferation Treaty effectively defines, is not one that Russia can prudently put aside based on an asserted technicality. This commercial nuclear transfer bears directly on the international community's ability to sustain a standard that is critical to preventing nuclear weapon proliferation.

The third issue where national policies will define the connections between nuclear weapons and nuclear energy involves *countries of particular concern*. When dealing with particular cases and regional realities, the sides will be drawn over how flexible to be with respect to the international nonproliferation regime. For example, after the administration of President William J. Clinton negotiated the Agreed Framework with North Korea and rewarded it with two light-water reactors, it was subject to a fair amount of criticism, at least initially, both domestically and from Europe because North Korea had clearly violated its Non-Proliferation Treaty and International Atomic Energy Agency safeguards obligations. There was concern both that the agreement would undercut the treaty and the agency and that the light-water reactors would ultimately be a source of plutonium if North Korea again chose to ignore its treaty commitments. These were not trivial concerns. However, in the context of the threat that the existing gas graphite reactor and reprocessing facility posed, the additional reactors under construction, and the limited options available for dealing with the situation, the administration made the judgment that, on balance, the deal was a good one. It was preferable to having North Korea walk out of the agency and treaty before the 1995 Review Conference and begin to accumulate approximately 150 kilograms of unsafeguarded, separated plutonium each year. However, the framework required that the International Atomic Energy Agency and the international community be flexible and wait to achieve their safeguards objectives until construction of the first light-water reactor advances to the point that items controlled by the Nuclear Suppliers Guidelines Trigger List are to be delivered. It is by then that North Korea must have come into compliance with its safeguards obligations. When that will happen cannot be predicted.

Would conventionally fueled plants have been preferable to light-water reactors in the deal from virtually every perspective? The answer is absolutely yes, except for the fact that North Korea would not have made the deal absent light-water reactors. Should the United States try to make the deal now? Certainly, as long as its allies, South Korea and Japan, concur and the move is made as a proposal to North Korea and not as evidence of America's unwillingness to stick with the framework as negotiated. The key here is to adopt the necessary flexibility to address the proliferation problem in its regional context and to use the international regimes to promote solutions rather than to block them. It seems extremely unlikely that the United States will succeed at defusing a very dangerous situation on the subcontinent if it approaches India and Pakistan armed only with the universal elements of the international regimes. The challenge is to find policy prescriptions that make sense in light of regional realities without undermining international norms. Doing so will not be easy in South Asia and, when the time comes, will be no easier in the Middle East. If the United States fails to find these formulae, it will cement the connections between nuclear power and nuclear weapons rather than loosen them.

The fourth and final issue has to do with *terrorism and nuclear weapons*. Many people do not see the two as linked except in made-for-television movies. This view is incorrect, even though nuclear weapons are hard to manufacture. It does not appear that there is a terrorist group on earth capable of assembling a nuclear weapon using an implosion system, even if it were given a sufficient quantity of plutonium and the help of a few refugees from nuclear weapon laboratories. Highly enriched uranium in a gun-type device might be a different story, although perhaps not. The concern derives from what could plausibly happen if fissile material became available to a country whose politics are radical and whose leadership had risk propensities much greater than those of the United States. A nation such as Iraq fits that description and has the capability to manufacture a simple fission weapon. It lacks only fissile material. If over time the Russian government fails to secure its fissile material, or if in the future China pursues a large nuclear program with significant use of plutonium fuels and less than adequate security in its civilian energy sector, or if some other country's plutonium is not sufficiently protected down to the few kilograms necessary for a weapon, Iraq could plausibly and secretly acquire that material. Although some might believe that Iraq would be deterred from even the threat of using nuclear weapons, it might not be deterred from transferring fabricated weapons to a terrorist group. Such a group could threaten and even detonate a weapon, delivering it to a port city in a container ship rather than on the end of a missile. The United States opens thirteen million containers each year. What would be the defense against, or even the deterrent to, such a threat? There is none.

The potential connection between nuclear power and nuclear weapons is real, and very troubling.

Nuclear Power and Proliferation

Richard Rhodes

Energy policy is not a zero-sum game, and this country would do well to continue encouraging conservation and efficiency and developing a variety of energy supplies—gas, renewables, and nuclear—while moving as rapidly as possible away from coal burning, its major source of electricity generation and also its most polluting. This chapter explores the nuclear power option, reviewing its history and present situation and its relationship to proliferation.[1]

The Current Status of Nuclear Power

It is instructive to look at the predictions about nuclear power against the current situation.[2] In 1975, nuclear power enthusiasts expected to see 450–800 power reactors in the United States by 2000. In fact, in that year there were 103, generating about 20 percent of U.S. electricity. Lovins has called this shortfall "the greatest collapse of any enterprise in the industrial history of the world."[3] It is unclear, however, why he refers to an enterprise that increased its output by more than 20 percent between 1990 and 2000—the equivalent of twenty-three new, 1,000-megawatt power plants—as "collapsed." Also in 1975, Lovins made what he called a "realistic" estimate that by 2000 direct and indirect solar power would meet about 39 percent of U.S. primary energy needs. By 2000, however, the actual figure for electricity generation by *all* U.S. renewables was only 2.3 percent.

These numbers illustrate not collapse for nuclear power or for renewables, but rather the growing pains of emergent new technologies. Initially nuclear power was oversold, and now renewables have been. Conservation and efficiencies intervened to slow demand, but even so U.S. electrical energy usage still grew faster than total energy from 1998 through 2000: by 4 percent from 1998 to 1999 and by 5 percent in 2000. That is a worldwide trend, with the share of final energy supplied by electricity growing rapidly and showing no sign of saturation.[4] Nor should it be overlooked that at least a third of U.S. efficiency improvements are actually the result of structural shifts within the economy, which has been moving from heavy manufacturing to service industries and imports of raw goods with embedded energy values for finishing.[5] In other words, the United States has transferred some of its energy usage offshore. Subsidized energy conservation has in fact been about twice as expensive as generated power.[6]

Government subsidies for energy technologies offer an interesting parallel. Nuclear power was heavily subsidized in its early years, as is common with new technologies. Nuclear facilities, by generating electricity more economically and with less pollution than any other source except large hydro, have repaid the nation several times over for its early public investment in research and development. In 1997 the U.S. research and development budget for commercial nuclear power was essentially zero. Since then it has increased, but it is still under $75 million per year. That is spare change compared with the government subsidies for renewables. The federal government subsidy for producers of green energy is 1.7 cents per kilowatt-hour, and many state and local governments add to that subsidy—in California's case, another 1.25 cents per kilowatt-hour. Subsidies for other renewables include double-declining depreciation, tax abatements, green energy surcharges on utility bills, and no escrow for demolition and disposal. In 1997 the U.S. federal research and development investment per thousand kilowatt-hours for coal and nuclear combined was 5 cents, for oil 58 cents, and for gas 41 cents. For wind, however, it was $4,769 per thousand kilowatt-hours and for photovoltaics, $17,006.

Despite these massive subsidies, amounting to some $30 billion to $40 billion in cumulative investment over the past twenty years, renewables are still significantly more expensive than nuclear or fossil-fuel electricity. Perhaps for that reason, production from renewables actually declined by 9.4 percent from 1997 to 1998. Evidently renewables have suffered growing pains even worse than those nuclear power experienced in its early days.

Consider, in addition, the unit capacity factor—the fraction of a power plant's capacity that it actually generates. The capacity factor for nuclear power was a source of amusement and condemnation back in 1980, when it averaged only 58 percent. By 2000, however, it was up to 87 percent, and so far this year it has been above 90 percent. By comparison, the Wisconsin Wind Power project, to take one example, generated just 13.2 percent of capacity in August 1999, and it produced at peak capacity for only about fifteen minutes on two days that month. Its annual average capacity has been 24 percent, with its lowest performance coming during the summer months, when electricity is most needed for air conditioning. Solar and wind installations are inherently low capacity because their fuels, wind and sunlight, come and go. They cannot be expected to improve their capacity much with increased operating experience, as nuclear has done.

Most air pollution in this country, including greenhouse gases, comes from coal burning and transportation. Coal burning also releases a hundred times as much radioactivity into the environment, megawatt for megawatt, as nuclear power does, because coal contains radioactive uranium and thorium and coal mining releases trapped radon. The Harvard School of Public Health estimates that air pollution from coal burning kills fifteen thousand people every year in the United States alone. Other estimates go as high as thirty thousand people every year. Although coal is cheap, it is also deadly. Nuclear power, which could replace coal with improved efficiencies and conservation, is nearly as cheap but without the air pollution. Between 1973 and 1999 U.S. nuclear power plants avoided 32 million tons of nitrous oxide pollution, 62 million tons of sulfur oxide pollution, and more than 2.5 *billion* tons of carbon pollution. Improved efficiency at

U.S. nuclear power plants has accounted for almost *half* of all industry carbon reductions. Even assuming that ten thousand or a hundred thousand years hence there is a significant cancer risk from buried nuclear waste leaking into the environment (an unlikely eventuality), how does that risk measure up against fifteen thousand or thirty thousand deaths *a year* from coal pollution? How does it measure up against the lead-pipe cinch that cancer will be preventable or curable in ten thousand years, presuming the human species is still on this planet as carbon-based life?

Nuclear Waste Disposal: A Political Problem

Nuclear waste disposal is a political, not a technical, problem. Waste disposal experts from twenty countries agreed collectively back in 1985 that disposal of nuclear waste could be done safely using available technology.[7] Because fission is six orders of magnitude more energetic than chemical burning, the volume of waste is small—some 3,000 cubic meters annually from all the operating nuclear plants in the world, compared with 50 million cubic meters of solid toxic waste from other industries in the United States alone. Spent fuel is highly radioactive at first, but 99.9995 percent of that radioactivity will have decayed away after five hundred years, leaving material that has no more radiotoxicity than a high-grade uranium ore deposit. Uranium oxide pellets are about as soluble as granite; to dissolve them completely would take ten million billion years.

Historically, the debate about nuclear power, which is usually cast as a technological one, has been and is a tacit debate about political control and social organization. In 1989 a proponent of nuclear power, David Fishlock, the science editor of the London *Financial Times*, characterized the debate in these words:

> Nuclear energy is controversial because a sector of Western society, with political ambitions to revert from an industrialized to an agrarian and craft-based society, sees nuclear technology as the most potent force for stability in our present way of life. The "nuclear threat" is nothing more than its ability to deliver low-cost, dependable power . . .

> [This] is an intolerable obstacle to [the Green Movement's] efforts to overturn industrialized society.[8]

From a different perspective, in 1975 Lovins wrote in a book he coauthored, *Non-Nuclear Futures: The Case for an Ethical Energy Strategy*:

> Fundamental to any discussion of energy alternatives is a choice—usually tacit but nonetheless real—of personal values. The values that make a high-energy society work are all too apparent today. The values that could make a lower-energy society work are not new; they are in the societal attic, and could be dusted off and recycled. They include thrift, simplicity, diversity, neighborliness, craftsmanship and humility.[9]

It would be nice to have the best of both these worlds, and, in truth, efficiency, conservation, and nuclear energy are not incompatible. Recently, however, Alan D. Pasternak of Lawrence Livermore National Laboratory looked into the relationship across the world between measures of human well-being and consumption of energy and electricity and found a disturbing correlation (figure 1).[10] He correlated the UN Human Development Index against annual per capita electricity consumption for sixty countries comprising 90 percent of the world's population. He found that the Human Development Index reached a maximum value when electricity consumption was about 4,000 kilowatt-hours per person. That is well below consumption levels for most developed countries—Japan comes in at 8,000 kilowatt-hours per capita, the United States at 13,000, and Canada at nearly 16,000—but well above the level for developing countries. The 4,000-kilowatt-hour threshold quantifies the bottom line for efficiency and conservation in developed countries. It also quantifies a much greater potential need for electricity in the developing world than that estimated by the developed world: That need is 102 percent more than the projections of the U.S. Department of Energy by 2020 under a scenario of low economic growth, and 52 percent more under one of high economic growth. Discrepancies in human development driven by poverty are measures of structural violence, which is ulti-

Figure 1. UN Human Development Index and Electricity Use, Various Countries, 1997

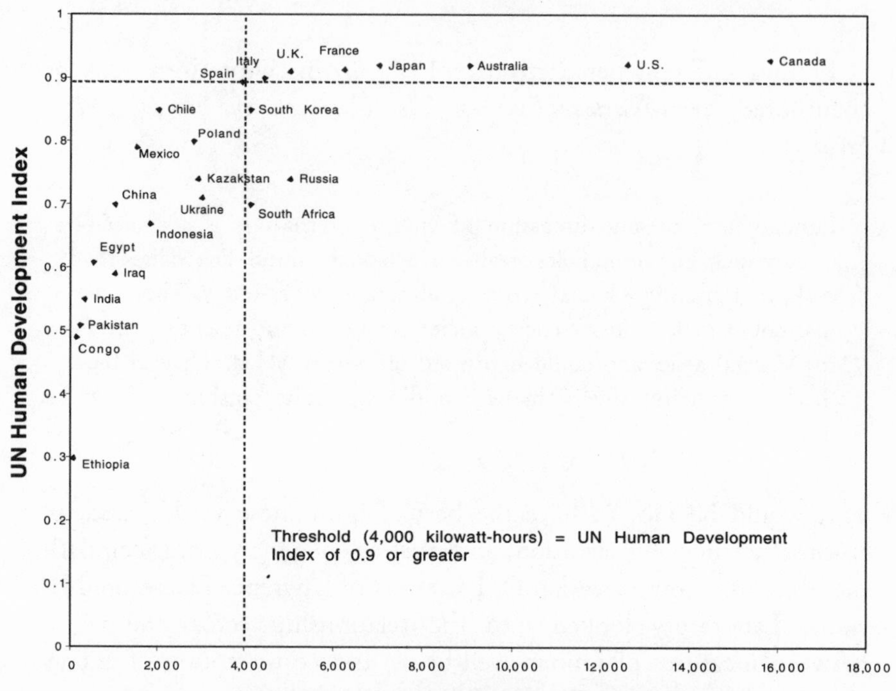

Source: Private communication to Richard Rhodes from Alan D. Pasternak, Lawrence Livermore National Laboratory.

mately the cause of social conflict within and between nation states (figure 2). How can the world move toward equity—toward significantly reducing structural violence—while controlling, much less reducing, greenhouse gases, without developing all the low-polluting energy resources it has (there is no such thing as a nonpolluting energy source)? Squabbling over which low-polluting energy source is greener in a world where two billion people have no electricity at all, where life expectancy increases directly with per capita gross national product, seems both elitist and immoral. It is obscene.

Dealing with Nuclear Proliferation

It is obvious that nuclear materials need to be policed, controlled, and accounted for. Even with that stipulation, proliferation must be

Figure 2. The "Economic Law of Life": Life Expectancy / Per Capita Gross National Product

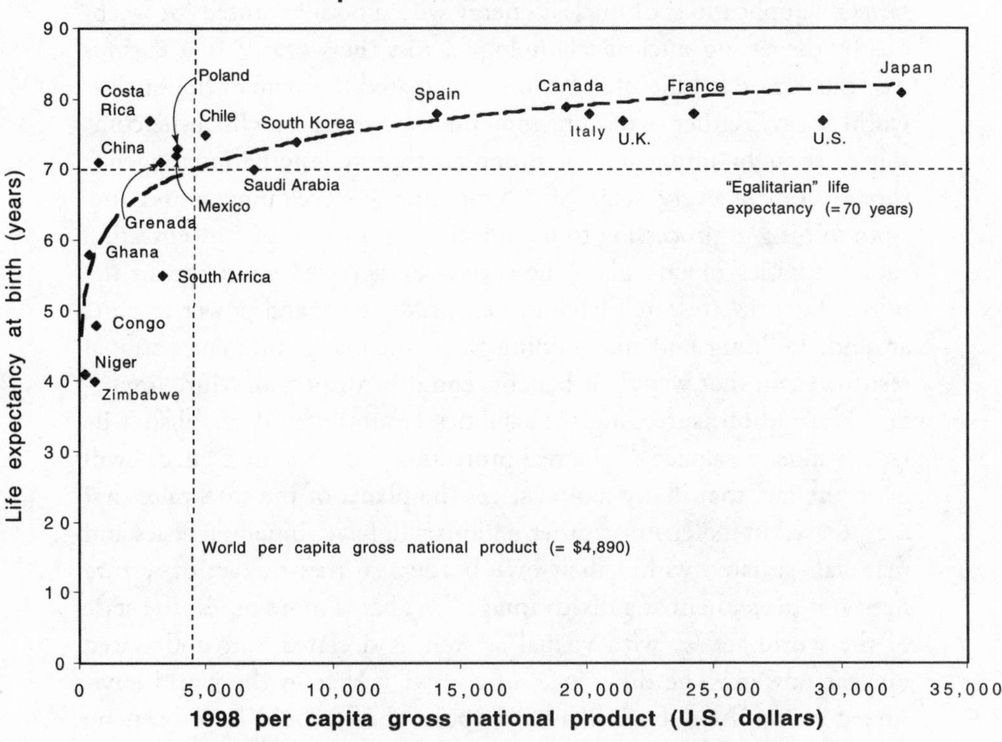

Source: World Factbook (Washington, D.C.: Central Intelligence Agency, July 1, 2001).

viewed as a political, not a technological, problem. No nation has gone nuclear with power reactor plutonium, nor is it clear why one would want to. If a nation, group, or individual wanted nuclear materials, there are better, faster, surer, cheaper, and secret alternative means to proliferation. No nation has ratified the Nuclear Non-Proliferation Treaty as a nonweapon state and then proceeded to become a weapon state. Weapons proliferation has decreased as nuclear power proliferation has increased, although the correlation is not significant. Eliminating all the nuclear power operations in the world would not prevent proliferation. Doing so might even encourage it by increasing structural violence.

As long ago as 1946, in the Acheson-Lilienthal Report that became the basis for the Baruch Plan for the international control of atomic energy, Robert Oppenheimer and the three industrialists who served as the report's board of consultants envisioned an arrangement where peaceful applications of nuclear energy would be substituted for weapons by dispersing nuclear technology across the world.[11] In a curious way the Acheson-Lilienthal Report anticipated the state of the nuclear world today. Rather than outlawing nuclear weapons, which the committee thought impossible, it proposed that an international agency should control every stage of atomic energy development and use, from mining to processing to manufacturing. Instead of concentrating these activities in one place, however, the agency would spread the mines, factories, research laboratories, production, and power reactors around, building and maintaining them on many different national territories. In that way their benefits could be dispersed. More important and radical, spreading the facilities around would establish a de facto strategic balance. "The real protection," the report argued, "will lie in the fact that if any nation seizes the plants or the stockpiles that are situated in its territory, other nations will have similar facilities and materials situated within their own borders so that the act of seizure need not place them at a disadvantage."[12] That is more or less the state of the world today, with virtual as well as declared and undeclared nuclear powers. The difference, of course, is that in the world envisioned in the Acheson-Lilienthal Report, there would have been no weapon stockpiles, so that if the system broke down, the delivery time of weapons to targets would be measured in months rather than minutes or hours.

Nevertheless, the report still offers a framework for a workable, peaceful nuclear future. Indeed, it seems that the only way to achieve a world free of the threat of nuclear weapons is, paradoxically, to disperse nuclear technology—nuclear power—widely across the world, because knowledge of such technology would always be the minimum necessary precondition to reconstituting national arsenals if one or more nations went rogue. *Having* that technology and *proliferating* with it are demonstrably different animals. The great majority of nations have chosen not to develop nuclear weapons despite having the technical capability and the infrastructure with which to do so.

Nuclear weapons do exist, as do the plutonium from those weapons, the plutonium in the spent fuel, and the plutonium that will be bred in the fuel now in use or yet to be loaded. Nuclear proponents and opponents alike understand that this material is the crux of the problem: So long as there is plutonium, there can be nuclear weapons, and generating nuclear energy breeds plutonium.

It follows that the only solution is to burn plutonium. Remove it from the weapons, lock it up, deform the pits, convert it into mixed-oxide fuels, and even breed it when that opportunity arrives. One way or another and systematically, under international supervision, burn it and eliminate it continuously from the world. Burn it and dispose of fifty years' accumulation of weapons material. Burn it and forestall its proliferation. Burn it and reduce greenhouse gas emissions. Burn it and reduce the cost of nuclear waste depositories. Burn it, and nuclear power becomes not a threat of proliferation but a guardian of world security.

A proposal that three Los Alamos National Laboratory scientists have developed suggests a way to do that. They envision a "comprehensive global nuclear materials management regime or system" based on the International Monitored Retrievable Storage Concept of "secure, retrievable storage of nuclear materials in a number of consolidated, internationally-controlled sites."[13] Their international regime would go beyond storage, however, to develop advanced proliferation-resistant plutonium separation that would then fuel special reactor- or accelerator-based systems for inventory reduction. "The process would produce significant amounts of electricity," they write. ". . . No commerce in plutonium among scattered nuclear facilities and reactors would occur. No transport of materials containing significant amounts of plutonium or other actinides of concern would take place once spent reactor fuel had been shipped to the international facility. . . . Such a management strategy would provide a robust back end for the nuclear fuel cycle that would deal with legacy materials."[14] Niels Bohr liked to say that "every great and deep difficulty bears in itself its own solution." Generating peaceful power from holocaustal weapons is a complementarity of which Bohr himself would be proud.[15]

Cesare Marchetti recently updated Nebojsa Nakicenovic's remarkable 1984 graph plotting the historical evolution of the world's primary energy mix as market fractions in tons of coal-equivalent (figure 3).[16] The dominant primary sources—wood, coal, oil, natural gas, and nuclear—rise and fall on Marchetti's graph, overlapped across time

Figure 3. World Primary Energy Substitution, 1860–1997

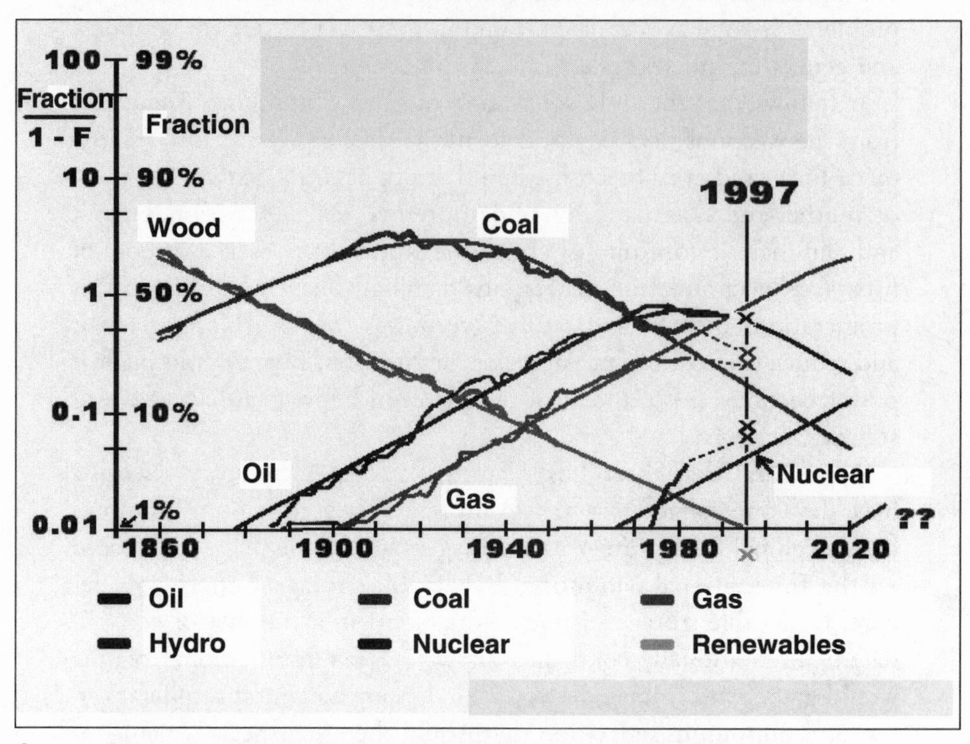

Source: Personal communication from Cesare Marchetti to Denis Beller, Los Alamos National Laboratory.

like a range of weathered mountains. Wood, which provided more than 50 percent of the world's primary energy in 1860, has declined to less than 10 percent today. Coal peaked around 1930. Oil, which began its ascent late in the nineteenth century, peaked at less than 50 percent around 1980. Natural gas, entering the market around 1900, ascends toward a peak of perhaps 60 percent in the first half of the twenty-first century. Marchetti's data demonstrate, in the words of one analyst, that "world-shaking events such as wars, skyrocketing energy prices, and depression had little effect on the overall trends."[17]

Nuclear power's share of the world's primary energy mix has increased in parallel with natural gas, displaced seventy years farther along the timeline. Nuclear grew too fast, slowed, and is poised to grow again. Five existing U.S. nuclear power plants have had their licenses renewed for twenty more years. Thirty more have filed for renewal, and most of today's existing U.S. plants will probably have their licenses renewed. Several American utilities are presently exploring finishing some of the plants left unfinished by the decline in electricity demand in the 1980s. This year Exelon filed for a Nuclear Regulatory Commission license for a South African–designed 110-megawatt-electric pebble-bed modular reactor that is inherently safe. It uses inert helium as its working fluid, passing the helium directly into a gas turbine heat engine. Exelon intends to build this fourth-generation reactor as a merchant power plant. That means it intends to put its own capital at risk to build it with the expectation that it can rapidly repay its expenditures by selling power at the market price. It expects to generate electricity that is cheaper than combined-cycle gas. Around the world, thirty-eight new reactors are under construction, eight of them in China and six in South Korea, with the latter planning eight more by 2015. Marchetti implies dominance for nuclear power in the second half of the twenty-first century. More significantly, no other primary source of energy has emerged to compete. Worldwide, renewables have not even reached the graph's 1 percent threshold.

James Lovelock, the English scientist who proposed the Gaia thesis, wrote recently of the "great benefits and small risks of nuclear power."[18] He noted, "Life began nearly four billion years ago under conditions of radioactivity far more intense than those that trouble the minds of certain present-day environmentalists. . . . I hope that it is not too late for the world to emulate France and make nuclear power our principal source of energy. There is at present no other safe, practical and economic substitute for the dangerous practice of burning carbon fuels."[19]

In short, nuclear power is alive and well.

Why Nuclear Power's Failure in the Marketplace Is Irreversible (Fortunately for Nonproliferation and Climate Protection)

Amory B. Lovins

For a decade or two, nuclear power has been dying a rather lingering death from an incurable attack of market forces. Nuclear salesmen are scouring the world, struggling to get a single order, while combined-cycle gas, wind, photovoltaics, and efficiency are struggling with more orders than they can handle. Moreover, the competitors are starting to pull ahead quite remarkably.

World renewables, far from being under 1 percent of primary energy production, as Richard Rhodes suggests, are around 20 percent by any standard industry database if traditional biofuels are included, and around 9 percent if they are not included. In this country, nuclear power has essentially the same primary energy output in a normal hydro year as renewables but produces almost twice the kilowatt-hours from the same generating capacity. However, when looking at the average annual growth rates in global capacity in the 1990s, a very different story emerges: nuclear, 1 percent; photovoltaics, 17 percent; and wind, 24 percent. Some might say that their growth is from a small base, but interestingly enough, in the past few years wind has added more gigawatts than there were nuclear starts, on average, during each year of the 1990s. That is actual gigawatts, not percentages. The reasons are not hard to find.

Because here the interest is in *delivered* electricity—the point at which it can actually be used—for remote sources, it is necessary to add an average embedded delivery cost. Currently in the United States that cost is about 2.7 cents per kilowatt-hour, with a range of roughly 1 cent for industrial to roughly 4 or 5 cents for residential. The marginal costs can be twice as high. Thus traditional nuclear technology delivers at about 10 to 15 cents or more, of which 4 cents to 7-plus cents is the delivered short-run marginal cost of operation, leaving out major repairs. A hypothetical pebble-bed modular reactor at 3.2 cents busbar would deliver at about 6 cents, and a new coal plant, at about 6 or 8 cents. A couple of years ago combined-cycle gas under constant-price thirty-year gas contracts cost 5 or 6 cents delivered, although temporarily the cost has blipped up about a cent because of a shortage of turbines, the result of its popularity. The remote wind turbines installed a couple of years ago delivered at 6 or 7 cents; the next generation now being ordered delivers at about 5 cents.

All of these remote sources have to compete with on-site resources that avoid the delivery cost. Photovoltaics are the most expensive currently but already are becoming competitive when integrated into building design. Their real cost dropped 43 percent in the 1990s and is continuing right down the experience curve that Robert Williams and others have documented. A couple of options achieve a thermal credit and hence have an even lower electricity cost net of that credit: less than 1 cent to 5 cents delivered for microturbine trigeneration at 90-odd-percent system efficiency, again with constant price gas; and industrial cogeneration, costing less than 1–2 cents. Finally, the cost of end-use efficiency ranges from negative to about 1 cent for most programs. Rhodes's numbers cannot have come from any of the standard literature.

Thus there are at least three abundant resources—efficient end-use, efficiently used gas (especially when thermally integrated), and wind power—all of which easily beat new nuclear plants in terms of the cost of delivered power, and many of which actually beat just the *operating* cost of old ones. It is not surprising that these systems are very popular in the market. Natural gas also turns out to be a rather ubiquitous and abundant resource, with a couple of hundred years' worth known. Rhodes's claim that renewables are in decline is readily contradicted by a reading of any of the industry literature. Look only at the cover story in the current issue of the European edition of *Fortune*, for example, on how renewables are the fastest-growing source in Europe. The case for renewables is even stronger in developing countries.

Any one of these three options would make nuclear power unnecessary and uneconomic. There are some other options that are not quite as big but that collectively are also important. Fuel cells and photovoltaics will raise that three to four or five options. Looking at "distributed benefits" seals the argument, as they increase the economic value by an order of magnitude for decentralized sources, as discussed later.

The focus here is, however, on end-use efficiency, the cheapest of these options, which Rhodes claims contributes "only marginally" to U.S. energy supplies. Even an update of the graph that appeared in *Foreign Affairs* twenty-five years ago (figure 4) shows that U.S. energy intensity is already down 40 percent from the projections that governments and industry made at the time, so that total primary energy consumption is now within a few percent of this author's "soft energy path" graph.[1]

Rhodes is correct that renewables are lagging behind the growth this author projected in 1976—because it specifically assumed a supportive, and not a largely hostile, policy environment. At the same time, it often goes unnoted how big the efficiency resource is. Over the past quarter century reduced energy intensity has become the nation's largest energy supply. It is more than five times as big as domestic oil output in this country, more than three times the size of oil imports, and more than thirteen times Persian Gulf oil imports. It is the fastest growing source, and most of the intensity reduction is the result of technical efficiency. The United States has doubled its oil productivity in the past quarter century and barely scratched the surface of the efficiency that is available and worth buying.[2]

Figure 4. U.S. Energy Consumption, 1975–2025: "Hard" and "Soft" Paths (1976) vs. Actual (1975–2000)

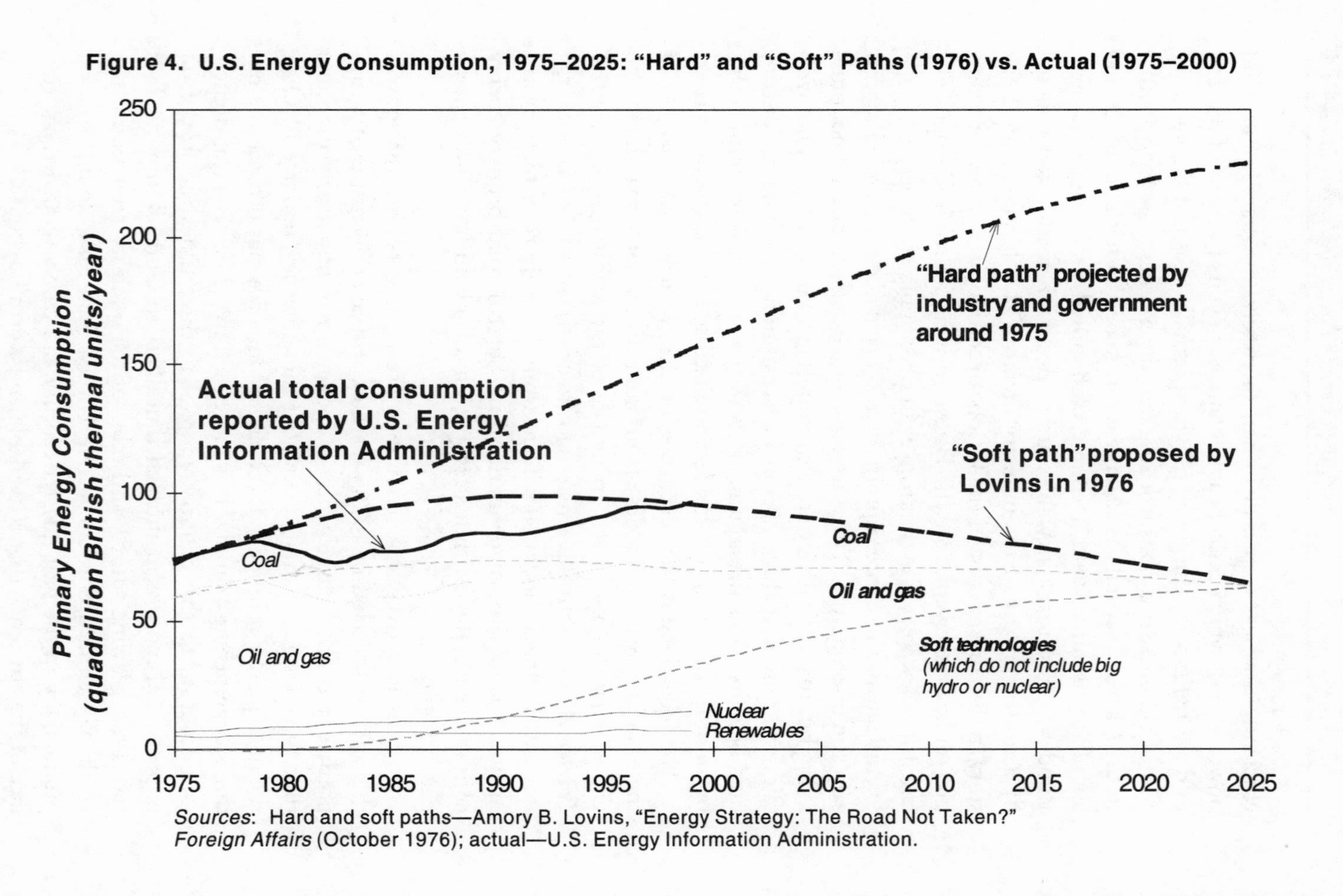

Sources: Hard and soft paths—Amory B. Lovins, "Energy Strategy: The Road Not Taken?" *Foreign Affairs* (October 1976); actual—U.S. Energy Information Administration.

The last golden age of energy efficiency was from 1979 to 1986, when the economy grew 20 percent while primary energy consumption fell 5 percent. That led to the 1986 price crash (which will recur later in this decade if current federal policy becomes law and is actually followed) and a decade of stagnation (although Californians saved over 10 gigawatts by the early 1990s before their attention wandered). Something odd has happened in the past few years, however. Starting in 1996, the United States has nearly regained the speed of savings that it had from 1979 to 1986. It has cut energy intensity 3.2 percent a year on average during 1996 through 2000, despite record low and falling energy prices through 1999. Some of this reduced intensity is the result of structural change, but many things other than energy price drive the important technical gains in end-use efficiency, and the savings are getting bigger and cheaper.

Of course, electrical savings are the most lucrative kind because each cent per kilowatt-hour is equivalent in heat content to oil at $17 per barrel. In fact, since 1995 electric intensity has been steadily falling in this country—in the past three years at an average rate of 1.6 percent a year—which had not happened before. There is still a vast unbought efficiency potential, enough to save upwards of three-quarters of electricity use in this country at less than short-run marginal supply costs. That unbought efficiency is a resource about four times the present nuclear output, and much cheaper than its operating cost. There is about a two-meter long technical bookshelf available on this subject for those interested in it.[3] There is also now a good understanding of the sixty to eighty market failures in buying efficiency and how to turn each one into a business opportunity[4]—partly to capture side benefits that are often worth an order of magnitude more than the energy savings themselves.[5]

Efficiency can also work very quickly, as it did in Southern California in the early to mid-1980s, when the ten-year forecast of peak demand was being reduced by about 8.5 percent *per year*, at about 1 percent of the long-run marginal cost of supply. There are other interesting examples. One of the more piquant occurred in 1990, when Pacific Gas and Electric signed up a quarter of new commercial construction in its territory for design improvements in three months. Finding that too easy, it raised the target for the next year—and they achieved it in the first nine days of January. Moreover, they did so with old delivery methods. New ones are even better: They not only market negawatts, but also make markets *in* negawatts and thus maximize competition in achieving them. The reason people tend to do end-use efficiency is that it is extremely cost-effective (at least when they are not too severely penalized for doing it, as they currently are in forty-nine states by utility regulations).

An article by Rhodes and Beller[6] states that end-use efficiency "remains stubbornly uncompetitive." That is contrary to a vast professional literature that is quite sophisticated, well-evaluated, and rigorously measured. That literature also shows that the costs and savings are accurately predictable and predicted. The historic average cost for all U.S. electric efficiency, to the utilities that produced it, was about 2 cents per kilowatt-hour. The figure was substantially lower than that in good programs, particularly those emphasizing the business sectors, because savings there tend to cost less than insulating houses. For example, a review of over two hundred programs by fifty-eight utilities found dozens of programs costing 0.4 cent to 1.1 cents per kilowatt-hour saved in 1988, and many of those were around a half cent or less per kilowatt-hour, with transaction costs on the order of 1 percent of the tariff.[7]

Those historic costs were obtained with old technologies and delivery methods. There are now much better technologies, many of which are now in volume production at competitive prices, and there are better delivery methods, better marketing, far better insight into the alchemy of turning market failures into business opportunities, much better customer awareness and eagerness, and continuing innovation that is expanding the technical potential for efficiency faster than it is used up.

The key headline, however, is the breakthrough in design integration. In the old days, say fourteen years ago, it was possible to do the sort of analysis presented in figure 5, which shows the supply curve of retrofittable U.S. electric efficiency. Around 1987, if fully installed wherever they would make sense, retrofits could save about three-quarters of U.S. electricity at an average 1986 cost of 0.6 cent a kilowatt-hour, based on the measured cost and performance of over one thousand technologies analyzed in some thousands of pages with many

Figure 5. A Preliminary Estimate of the Full Practical Potential for Retrofit Savings of U.S. Electricity at an Average Cost of ~0.6 Cent per Kilowatt-Hour

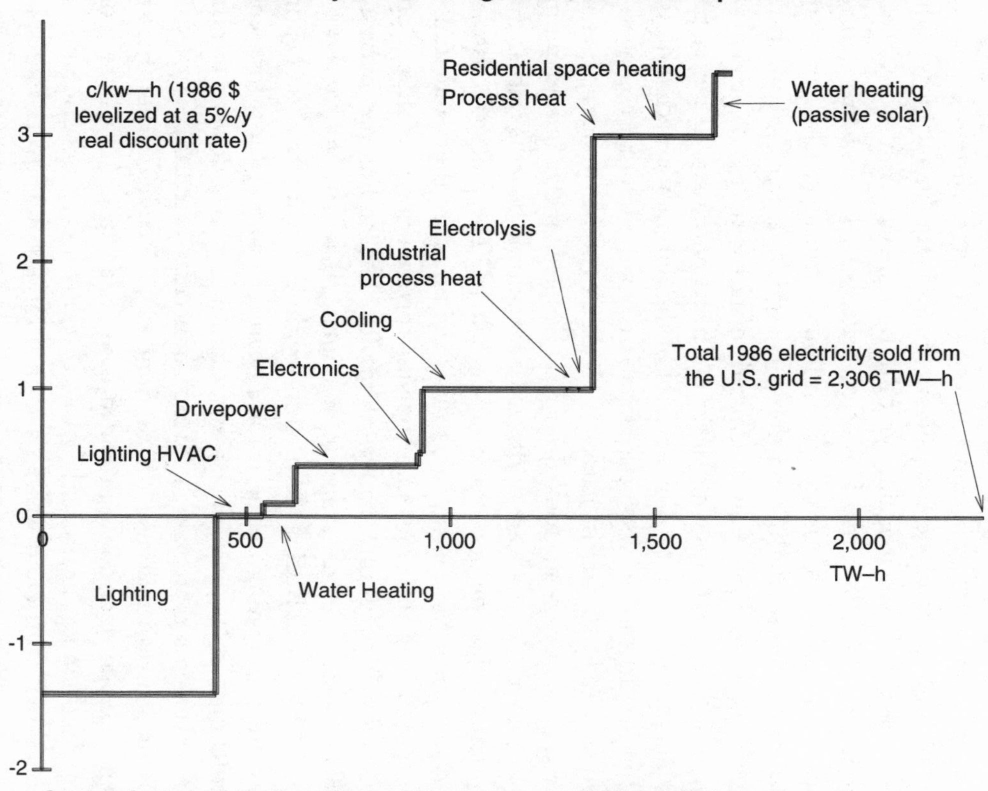

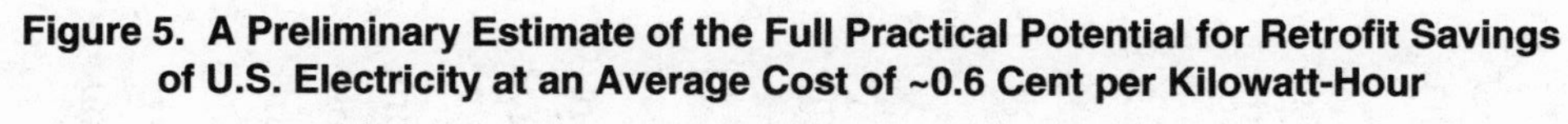

Source: Analysis of the Rocky Mountain Institute, based chiefly on the first editions of the Rocky Mountain Institute/Competitek *The State of the Art* series, 1987, republished as E SOURCE *Technology Atlas.*

footnotes per furlong. Every term shown in that "supply curve" of negawatts is now known to be conservative in both price and quantity.

Finally, there is the biggest improvement of the past fourteen years: namely, the discovery of how to escape from the diminishing returns idea that the more savings a person buys, the more steeply the marginal costs rise until it becomes too costly to buy. Current knowledge reveals how to combine components into systems in such a way that the cost comes *down* again, typically by achieving multiple benefits from single expenditures. This often makes very large savings cost *less* than small ones. That is why, for example, an industrial client of the author's recently cut the pumping energy in a standard industrial pumping loop, supposedly optimized to start with, by 92 percent—a twelvefold reduction—with lower construction costs and better performance in every way. This reversal of the normally expected diminishing returns is not rocket science. It is good Victorian engineering rediscovered. All that is needed is to use fat, short, straight pipes instead of skinny, long, crooked ones. The pipe diameter and the pumping system *as a whole* are optimized for multiple benefits, rather than optimizing just the pipe for a single benefit. The pipes are also laid out first, then the equipment, so that the pipes are short and straight instead of long and crooked.[8] Applying this thinking to buildings produces some significant empirical results. The author has harvested twenty-seven banana crops in a climate that can go to −44°C (−47°F) in the Rocky Mountains—without having to use a heating system. Eliminating the need for a heating system reduces total construction cost, because the superinsulation, superwindows, and other improvements that make a furnace unnecessary actually cost less to install than a furnace would cost. Similarly, a Pacific Gas and Electric experimental tract house provided comfort at 46°C (115°F) without air-conditioning, again with lower construction costs. The author's household electric bill was $5 a month for a 4,000-square-foot space (without taking credit for producing five to six times that much electricity with photovoltaics). The efficiency improvements alone yielded a 90 percent household electric savings and a ten-month payback with 1983 technology, and that payback also includes a 99 percent savings on water and space heating energy.

A new house was built a few years ago in steamy Bangkok with 90 percent less air-conditioning energy, better comfort, and no extra cost. In big office buildings the efficiency gains are typically a factor of four to ten compared with normal practice, yet the capital cost is typically several percent lower (the reduction occurring largely in the mechanical budget), the construction faster, and the human and market performance better. The author's firm showed a client how to retrofit a large Chicago curtainwall office tower to save three-quarters of its energy use at no more cost than the normal twenty-year renovation that saves nothing.[9] The consultancy's record so far is a retrofit it designed to save 97 percent of the energy used for cooling a California office with, again, better comfort and good cost-effectiveness.

Industrial examples illustrate a comparably impressive potential. Retrofitting thirty-five improvements in motor systems, which use three fifths of the world's electricity, can typically save about half the electricity with a 100–200 percent annual return on investment after taxes. The reason it is so cheap is that the first seven improvements provide another twenty-eight for free. Lately, similar returns on investment have been empirically shown with retrofits that save 50-odd percent of the energy used in microchip fabrication plants to produce and deliver chilled water and clean air. That is partly why the fourth biggest chip maker in the world, ST Microelectronics, is targeting zero net carbon emissions by 2010, and Dupont is planning to boost its energy productivity at least 6 percent a year in this decade, all in the name of shareholder value. Dow Louisiana did not even need techniques like that to get a return on investment of over 200 percent on simple energy-saving retrofits that added $110 million per year to its bottom line.

On the generation side, the modest amount of remaining electricity will have to come from somewhere. This is a question not only of source but also of scale—the right size for the job. It turns out that in the United States, for example, average loads in three quarters of households do not exceed 2.4 kilowatts, and three quarters of commercial customers have average loads of 10 kilowatts or less. Given those figures, does a plant producing hundreds or thousands of megawatts in one place make sense? Distributed generation has actually taken over the market, although rather quietly. Large combined-cycle gas plants in the hundreds of megawatts range are still being ordered, but they are already obsolete on the margin.

Distributed generation is especially valuable because it can be quick. In the mid-1980s utilities in California were being offered private generation averaging 12 megawatts per unit, mostly renewable, rising at a rate equivalent to a quarter of the total peak load *per year*. There were similarly impressive small power commitments in the northeast. By 1998, 38 percent of California's net electric generation was renewable, 56 percent in Maine, and 11 percent in the United States as a whole.

Contrary to Rhodes's claims, the land and materials needed for renewables are quite modest. Denmark is now one-sixth wind-powered, and it is on target for 50 percent in or before 2030. It has not had a land-use issue and does not expect to. It also solved the alleged intermittence problem a long time ago. Modern wind turbines occupying 5 percent of four Montana counties could make 20 percent of U.S. electricity, according to Battelle Pacific Northwest Laboratory. Rhodes's numbers on this point are off by a factor of roughly twelve. The same is true for photovoltaics: Total annual electric use in the United States could come from ordinary photovoltaics occupying about a half of a square one hundred by one hundred miles. In reality, actual installations would not be centralized like that; they would be spread out, sharing land with other uses, and the photovoltaics would generally be put on buildings and usually integrated *with* buildings—a well-developed innovation that often makes them cost-effective today.

Based on empirical evidence, the energy paybacks for these technologies are months to a few years. An interesting example that Williams worked out in 1985 and the author did so recently is that a kilogram of silicon in a thin-film photovoltaic can actually produce more electricity than a kilogram of uranium in a light-water reactor—because although the silicon does not undergo the very high-energy nuclear reaction that the uranium does, it is used over and over, not just once.

"Distributed benefits" really change the game.[10] There are more than a hundred hidden benefits of making electrical resources the right size for the job, and typically those benefits increase the economic value by an order of magnitude. The biggest gains are in financial economics, the next in electrical engineering, and there are a lot of miscellaneous effects, although externalities are not being counted here. The end result, however, is that photovoltaics are actually cost-effective right now in most places and uses if their benefits are properly counted.

Another game changer—one that comes from transportation technology—will flower in this decade. It can have a huge effect on distributed generation. A little company chaired by the author recently developed a concept car that illustrates what can be done with ultralight, ultralow drag, hybrid-electric design. This Hypercar™ design happens to be equivalent to a Ford Explorer or a Lexus RX-300—a very roomy and capable five-seat sports utility vehicle that can haul half a ton up a 44 percent grade. Because the car is made of carbon fiber, it can also run head-on into a wall at 35 miles an hour with no damage to the passenger compartment, or run head-on into a Ford Explorer over twice its weight, each traveling at 30 miles an hour, and save the occupants from serious injury. It accelerates from 0 to 60 miles per hour in 8.2 seconds and gets 99 miles per gallon–equivalent—quintupled efficiency. It drives 330 miles on seven and a half pounds of safely stored direct hydrogen running a fuel cell. Because the car is so efficient, it can cruise at 55 miles per hour on the energy that the Lexus uses just for its air conditioner. There is no emission except hot water (maybe a coffee machine in the dashboard?). The car is very sporty, extremely reliable, and flexible. It is software-dominated. It can be designed for a 200,000-mile warranty. The body does not rust or fatigue, and it bounces off a collision at 6 miles per hour with no damage. It can be produced at mid-volume at a competitive cost, using a tenth the normal amount of capital, assembly effort and space, parts, and (ultimately) product cycle time.

Why is this car relevant to electricity? This class of vehicles is not just a "nega-OPEC" in ultimate global oil-saving potential, or a zero-emission vehicle that decouples driving from climate and smog and gets the United States out of the Iron Age. It can also be designed as a power plant on wheels. When the car is parked, which is about 96 percent of the time, it can be plugged in as a distributed generator. A full U.S. fleet of such light vehicles would have a generating capacity of 5–10 terawatts. Running on neat hydrogen, the fuel cells can readily be built to run for decades extremely reliably, silently, and cleanly. The generating capacity is about six to twelve times as much as what all power companies now own. As such, it does not take many people liking this new value proposition—from which the owner earns back up to half the lease cost of his or her car by selling electricity at the real-time price when parked—to put the coal and nuclear plants out of business. This advance is happening quickly, partly because the author put the work into the public domain in 1993 and maximized competition in exploiting it. Rocky Mountain Institute has also published a profitable transitional path to a climate-safe hydrogen economy.[11] New nuclear plants have no place in that economy because they are not a competitive way to make hydrogen. Fuel cells simply put nuclear power out of business even faster.

Developing countries are going to move in exactly the same direction when they realize what their opportunities are. One of the more interesting examples is China, which has more than doubled its energy productivity and is making further major improvements (developing countries on average are only about one third as energy-efficient as Organisation for Economic Co-operation and Development countries to start with). China has cut its coal output by roughly a third in the past five years, and will soon do so by half, to boost economic development and public health. A very rapid shift to efficiency, gas, and renewables is underway, and China is very interested in hydrogen as well. It has announced a moratorium on nuclear orders for at least five years.[12]

Generically, countries in the south—the developing world—are saving energy and carbon at least as fast as the north in percentage and maybe in absolute terms. In general, they are doing so for development purposes, not the environment, but that is all right. The capital savings they get by increasing electricity's productivity rather than expanding the supply of electricity is three or four orders of magnitude—huge leverage for development in a world where a quarter of development capital goes to the power sector.

Why Nuclear Power Makes Global Warming Worse

The commercial collapse of nuclear power in favor of cheaper alternatives is good for climate protection. Assume, pessimistically, that saving a kilowatt-hour costs as much as 3 cents, while generating a new nuclear kilowatt-hour costs, optimistically, as little as 6 cents delivered—assuming a *really* cheap pebble-bed reactor. Then each 6 cents spent on such a nuclear kilowatt-hour could have been used to buy *two* efficiency kilowatt-hours instead. By buying the costlier instead of the cheaper option first, an unnecessary additional kilowatt-hour has been generated from, say, coal. Unless nuclear power is the cheapest way to meet energy service needs, buying it will actually worsen climate change. That is, the order of economic priority is also the order of environmental priority. Whether nuclear can beat coal does not matter because neither of them can beat other options that are free of carbon dioxide. It is quite true, as Rhodes says, that if fossil fuels had to pay for containing their emissions, they would cost more. However, that charge would competitively benefit not nuclear power so much as the still cheaper, faster, and more attractive alternatives such as efficiency and renewables that emit no carbon dioxide.

Nonproliferation Benefits

Finally, a least-cost and hence non-nuclear energy strategy will help nonproliferation as well. A paper published twenty-one years ago[13] may have provided the first rat-proof description of what an effective and internally consistent nonproliferation regime requires. The first necessary condition is the commercial collapse of nuclear power. Nuclear power is a very convenient route to bombs, although not technically ideal (in the sense that cabinetmakers would rather have teak than pine, but it is possible to make a perfectly good cabinet out of pine). It is convenient because it is innocent-looking, socially approved in many circles, and heavily subsidized.

A second necessary condition is the rise of clearly better and cheaper energy options.

A third condition, as Rhodes correctly says, is the end of the cold war and bipolar hegemony and of attempts to bully the world by having bombs.

Most readers twenty-one years ago would not have thought those three conditions were possible, but they all have come to pass. Thus their logic is still correct and merits revisiting. It says that in a world without nuclear power, all the ingredients needed for do-it-yourself bomb kits by any of the twenty or so known methods would no longer be ordinary items of commerce. Because their civilian cover would be gone, they would be harder to get, more conspicuous to try to get, and politically costlier for all parties to be caught trying to get. The ambiguity would be removed. The resultant transparency would smoke out proliferators and their suppliers and enable the world to focus intelligence resources on far fewer transactions. This does not, of course, make proliferation impossible, but it would be a lot harder—even, in most cases of practical interest, prohibitively difficult. Those wanting energy for development would have to explain why they are choosing the costliest way to get there, as it cannot be just for energy and economic reasons.

Certain political conditions are essential to make this approach to nonproliferation work. At their heart is the central *purpose* of the Nuclear Non-Proliferation Treaty bargain: to ensure fair access to affordable energy for development—but not specifically to *nuclear* energy now that there are better solutions. Nuclear experts negotiated the nuclear part of the Nuclear Non-Proliferation Treaty within a nuclear context. If *energy* experts were involved rather than nuclear experts, the outcome would be a much broader portfolio of solutions that make far more sense for a developing economy. That approach has never happened in nonproliferation's history. It is time that it did. By educating people on why bombs make the world less secure and bespeak national immaturity, and by setting an example, it is possible, by emphasizing the role of ritual and symbolism, to try seriously to kick the habit, get together at Hiroshima every year for meetings of Bombaholics Anonymous, have deep cuts, and build a new security triad based on conflict prevention or avoidance, conflict resolution, and nonprovocative defense. The result could actually be a more secure world at least cost.

In conclusion, somebody said that nuclear fission is "a fit technology for a wise, farseeing, and incorruptible people." The enormous devotion of talent, work, hope, and investment in this technology deserved better. It is one of the great tragedies of modern times, and it is still distorting the world's energy and security choices.

A basic lesson from the experience with nuclear technology is that shielding any technology from political and market accountability for long enough can result in really big mistakes. The best legacy from this tragedy would be to avoid making the same mistake all over again. Market discipline is a good substitute, at least a good start. It draws the right conclusion. Trying to ignore or reverse or delay that conclusion has a huge opportunity cost. It is time to accept market realities and to design an orderly terminal phase for nuclear power.

There is a greater likelihood that communities will accept nuclear waste if it is not an open-ended commitment to unlimited quantities, but rather an effort to deal responsibly with existing waste without the prospect of more. The nuclear religion that does not allow an admission that a terminal phase is underway is also the main barrier to public acceptance of nuclear waste.

The opportunity is at hand to turn nuclear power's commercial collapse and the rise of better energy alternatives into the long-awaited missing step toward effective nonproliferation.

6

Nuclear and Alternative Energy Supply Options for an Environmentally Constrained World: A Long-term Perspective

Robert H. Williams

Nuclear power is a commercial technology that offers the potential for providing electricity with zero emissions of air pollutants and greenhouse gases. Despite this promise, the nuclear power industry is stagnating. Most energy projections show that, although some new capacity will be added (primarily in Asia), there will be little or no net growth, or even a decline, in nuclear generating capacity worldwide over the next two decades.[1]

Nuclear power faces four serious challenges: (1) costs that are typically higher than those for alternatives; (2) concerns about reactor safety; (3) the lack of significant progress in dealing with the disposal of radioactive waste; and (4) the nuclear weapons connection to nuclear power. The recent World Energy Assessment[2] concluded that the prospects for addressing the reactor safety challenge satisfactorily are good and that the waste disposal problem can probably be solved technically—although it will be difficult to convince the public that the problem is soluble. The World Energy Assessment reached no judgment on the cost challenge ("the proof is in the pudding"). It expressed skepticism regarding the prospects for coping effectively with the nuclear weapons connection to nuclear power. The reason is that the challenge of separating the peaceful atom from the military atom is a formidable one. It is especially so at the high levels of nuclear power development needed to make a dent in climate change mitigation by serving as an alternative to continued reliance on fossil fuels over the longer term.

The Climate Change Mitigation Challenge under a "Business-as-Usual" Scenario

The climate change challenge is forcing policymakers to take a long-term perspective in energy planning (on the scale of a century). The reasons are twofold: the likelihood that radical technological change for energy is needed to deal effectively with climate change; and the fact that fifty years or more are needed to transform the energy system fundamentally.[3]

The IS92a global energy scenario (see table 1) of the Intergovernmental Panel on Climate Change,[4] often referred to as a "business-as-usual" scenario, represents a plausible course for global energy under a public policy that does not consider climate change concerns. IS92a has been widely used as a framework for understanding the climate change challenge and also provides a useful framework for understanding energy-related risks other than climate change—such as the nuclear weapons connection to nuclear power and land-use and other challenges posed by renewable energy options. Under IS92a, emissions of carbon dioxide from the burning of fossil fuels increase from 6.2 gigatonnes[5] of carbon in 1997 to 19.8 gigatonnes in 2100; cumulative emissions of carbon dioxide in the twenty-first century amount to 1,340 gigatonnes of carbon; and the atmospheric level of carbon diox-

Table 1. Global Energy

	Actual, 1997	*IS92a projection for 2100*
Electricity generation (terawatt-hours per year)		
Coal	4,818	15,480
Oil	1,244	531
Natural gas	2,246	915
Synthetic liquids/gases from coal	-	3,017
Hydroelectric	2,574	7,660
Wind	192	20,405
Photovoltaic		
Biomass		1,381
Nuclear	2,266	18,695
Subtotal	13,340	68,084
Carbon dioxide emissions, power sector (gigatonnes of carbon per year)	1.9	4.9
Fuels used directly (exajoules per year)		
Coal	43.2	132.7
Oil	142.7	94.6
Natural gas	63.0	37.3
Synthetic liquids/gases from coal	0	276.5
Synthetic liquids/gases from biomass	0	126.5
Subtotal	248.9	667.7
Carbon dioxide emissions, fuels used directly (gigatonnes of carbon per year)	4.3	14.9
Primary energy requirements (exajoules per year)		
Coal	97.9	718
Oil	156.9	100
Natural gas	88.5	47
Biomass	-	205
Total carbon dioxide emissions (gigatonnes of carbon per year)	6.2	19.8

Sources: Global data for 1997 are from U.S. Department of Energy, Energy Information Administration, "International Energy Outlook 2001," DOE/EIA-0484 (2001), Washington, D.C., March 2001. The IS92a projection is the "business as usual" global energy scenario presented in Intergovernmental Panel on Climate Change, *Climate Change 1994—Radiative Forcing of Climate Change and an Evaluation of the IPCC IS92 Scenarios* (Cambridge: Cambridge University Press, 1994). Energy quantities are presented on a higher heating value basis.

ide increases from the present 365 parts per million (by volume) to 700 parts per million by 2100.

Although there are many uncertainties regarding the potential impacts of such a rise in carbon dioxide in the atmosphere, the impacts are likely to be severe,[6] suggesting the importance of exploring whether it would be feasible to evolve an energy system for which carbon dioxide emissions are such that the atmosphere could be stabilized at 550 parts per million (double the pre-industrial level), or even 450 parts per million of carbon dioxide. Stabilizing the atmosphere at 550 parts per million would require reducing cumulative emissions from 2000 to 2100 by more than 500 gigatonnes of carbon relative to those projected in IS92a and evolving an energy system that emits no more than about 5 gigatonnes of carbon by 2100. Stabilizing at 450 parts per million would require reducing cumulative emissions from 2000 to 2100 by more than 850 gigatonnes of carbon relative to IS92a and evolving an energy system that emits about 3 gigatonnes of carbon per year by 2100.[7]

Examining the details of the IS92a projection illuminates the climate change mitigation challenge (see box 1). Under IS92a, the historical trend toward electrification of the energy economy continues, with the electricity share of secondary energy consumption increasing to 28 percent in 2100, nearly double the current share. Yet the power sector's share of carbon dioxide emissions declines from one-third in 1997 to one-quarter in 2100 at the global level (see figure 6). The declining share of emissions from the power sector arises in part because of the expectation that the contributions from noncarbon supplies will grow to 71 percent in 2100,[8] up from 38 percent in 1997.[9] Another important reason is an increase in the carbon intensity of the fuels used directly in the latter half of the century as a result of the expected peaking of conventional oil and natural gas production before the middle of this century (see box 1).

Nuclear Power in Climate Change Mitigation and Associated Nuclear Weapon Risks

Under IS92a, nuclear capacity grows from 350 gigawatts-electric in 1997 to 2,700 gigawatts in 2100. Even with this substantial growth in capacity, the nuclear contribution to climate change mitigation is relatively modest. If coal entirely replaced nuclear power in IS92a, emissions of carbon dioxide in 2100 would be 24 gigatonnes of carbon per year, about 20 percent more than in IS92a.

Box 1. The IS92a Scenario for Global Energy

Under IS92a, global population grows from 5.9 billion in 1997 to 11.3 billion in 2100, global gross domestic product grows at an average rate of 2.2 percent per year (so that gross domestic product per capita grows fivefold in the twenty-first century), while primary energy grows about 1 percent per year more slowly (approximately the historical rate of decline in energy intensity).

The trend toward electrification of the global energy economy continues under IS92a, with per capita electricity generation growing to 6,000 kilowatt-hours in 2100 (about one half the 1997 U.S. rate), up from 2,300 kilowatt-hours in 1997. Under IS92a, direct use of fuels (in applications other than for electricity generation) grows much more slowly: Direct secondary fuel use per capita grows less than 40 percent (to 59 gigajoules in 2100, about a quarter of the 1997 U.S. rate, up from 42 gigajoules in 1997).

Because of an expected peaking of global conventional oil and natural gas production before mid-century,* unconventional energy sources will be needed to provide fuels used directly. IS92a projects large roles for both coal- and biomass-derived synthetic fuels under business-as-usual conditions—with synthetics (69 percent derived from coal and 31 percent from biomass) accounting for three quarters of both liquid and gaseous fuels by 2100. Despite a projected large role for biomass (205 exajoules per year by 2100, see table 1), which has a carbon intensity of zero, the average carbon intensity of fuels used directly in 2100 is about 30 percent higher than in 1997.

* Although there is no imminent danger of running out of conventional oil and gas, productive capacity is expected to be constrained after about one half of remaining exploitable conventional resources have been used up, in large part as a result of the tendency to exploit the largest fields first.

Nuclear power could potentially play a larger role in climate change mitigation. For example, an increase in nuclear capacity at the end of the century to 4,900 gigawatts would be enough to displace *all* coal power. If such a nuclear-intensive variant of IS92a were to be realized, emissions of carbon dioxide in 2100 would be 16 gigatonnes of carbon per year, about 20 percent less than in IS92a. For this level to be reached, nuclear power plants would have to be built at an average rate[10] of 85–90 gigawatts per year during the century.

At this high a level of nuclear power development, the nuclear weapon link to nuclear power (the risks of both nuclear weapons proliferation by nation states and of terrorists acquiring nuclear weaponry) would come into sharp focus. Consider first the case where uranium resource constraints force a shift to conventional plutonium breeder reactors sometime during the second half of this century so

Figure 6. Distribution of Carbon Dioxide Emissions from Fossil Fuel Burning by Activity (percent)

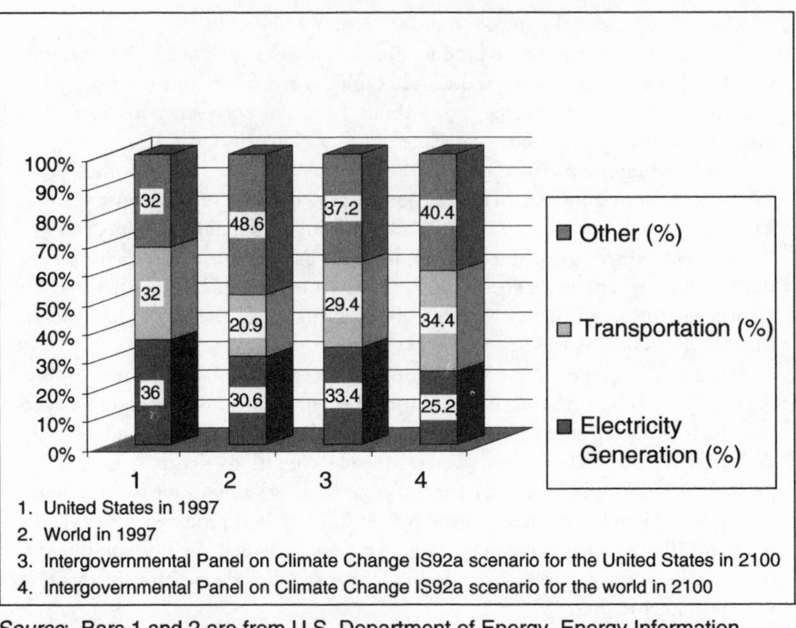

Source: Bars 1 and 2 are from U.S. Department of Energy, Energy Information Administration, "International Energy Outlook 2001," DOE/EIA-0484(2001), Washington, D.C., March 2001; and bars 3 and 4 are from Intergovernmental Panel on Climate Change, *Climate Change 1994—Radiative Forcing of Climate Change and an Evaluation of the IPCC IS92 Scenarios* (Cambridge University Press, 1994).

that these reactors are in wide use by 2100, along with fuel reprocessing and plutonium recycling. Under these circumstances each 1-gigawatt power plant would discharge in its spent fuel 1,000 kilograms of plutonium each year that would be recovered via reprocessing and used in fresh fuel. The amount of plutonium circulating in global commerce would be about 5 million kilograms per year. The nuclear weapon link to nuclear power in this scenario is apparent from the fact that the amount of plutonium needed to make a nuclear weapon is less than 10 kilograms.

The institutional challenges associated with preventing significant quantities of this plutonium from being diverted to weapon purposes are daunting. Attempts are being made to design technologies for which the energy benefits of the plutonium economy might be exploited without the risks implicit in the conventional plutonium economy.

One such technology is a metal-cooled fast reactor for which plutonium is never fully separated from the fission products.[11] Nevertheless, such systems would provide their operators with extensive knowledge, experience, and facilities for chemical processing of intensely radioactive spent fuel, all of which could provide the basis for moving quickly to separating plutonium for a weapon program, should a decision be made to do so.[12]

A modest-scale (100-megawatt-electric), compact[13] version with a lifetime (fifteen-year) sealed core has been proposed for application in developing countries as a way to avoid building a proliferation capacity.[14] The system is intended to be proliferation-resistant in that the reactor core would be sealed so that individual fuel assemblies could not be removed. The entire sealed core would be delivered as a unit to the power plant site and returned to the factory at the end of its useful life. However, the reactor core would contain 2.5 tonnes of plutonium, so that high security would have to be provided to deter the theft of sealed reactor cores during transport to and from deployment sites. If the technology were to catch on, providing adequate security would be a daunting challenge. Suppose one half of the 4,900 gigawatts of nuclear capacity in 2100 in the nuclear-intensive scenario discussed above were based on such reactors and that this capacity were built up at a linear rate beginning in 2040. During the period 2040 to 2100 the average rate of deployment would be about 1,000 reactors per year.[15]

If uranium could be extracted from seawater at a competitive cost,[16] a shift to a plutonium economy might be avoided altogether. In that case continued reliance on once-through nuclear fuel cycles that are inherently more proliferation-resistant than fuel cycles involving plutonium recycling might be possible. Given the large number of nuclear plants, however, proliferation concerns would still be considerable. Harold Feiveson has described a nuclear-intensive energy future in which the 100-megawatt pebble-bed modular reactor operated on a once-through fuel cycle[17] becomes the norm and modest-scale enrichment plants (each having a capacity of 500 tonnes of separative work units per year and serving about twenty-four such reactors) would be widely deployed.[18] About two thousand such enrichment plants would be needed to support 4,900 gigawatts of nuclear capacity. Assuming a critical mass of 15 kilograms, the enrichment requirement would be 3.4 tonnes of separative work units per bomb, so that each enrichment plant would have the capacity to make about a hundred and fifty bombs per year from natural uranium. More troubling still is that, for the 8-percent-enriched uranium prepared for the pebble-bed modular reactor, 84 percent of the enrichment required to make 90-percent-enriched uranium for a bomb has already been carried out. Thus, starting with 8-percent-enriched uranium, each enrichment plant could make 875 bombs per year. Feiveson concludes, "So, the bottom line? Lots of enriched uranium too close to bomb quality, lots of separation plants, and lots of incentive for innovation to make isotope separation cheaper and quicker. To me this is an unsettling prospect."[19]

Recently the author was involved in a major review of advanced fossil and nuclear energy technologies and in the process became sensitized to the shortcomings of alternative proliferation-resistant nuclear reactor and fuel cycle technologies.[20] Largely as a result of that experience the author is coming to the view that the nuclear weapon connection to nuclear power cannot be made acceptably low at levels of nuclear power development high enough to make a dent in addressing the climate change mitigation challenge, unless all sensitive facilities—enrichment plants, reactors, reprocessing plants, and fuel fabrication plants—are clustered in large, heavily guarded nuclear parks maintained under international control to reduce the risks of proliferation and diversion. There is no doubt that the nuclear park concept is technically feasible and would reduce the dangers of proliferation and diversion substantially. Much more questionable, however, is whether it is politically realistic to expect all the world's countries to place a major component of their electricity supplies under international control, and to agree on the administrative arrangements for doing so. Acceptance of international controls would be especially difficult for those countries that are world leaders in nuclear technology development and deployment and that see nuclear power as offering energy autarky.

The question remains, however, whether a climate change mitigation strategy that requires nuclear power to be developed at such high levels is worthwhile. Answering this question requires knowing the prospects for the alternatives to nuclear power in a world seeking to constrain emissions of greenhouse gases. The prospects for some leading supply alternatives to nuclear energy for mitigating the climate change challenge in the power sector are first discussed. Then major options are discussed for reducing emissions of carbon dioxide for fuels used directly, which account for three quarters of carbon dioxide emissions in 2100 under IS92a.

Achieving Deep Reductions of Carbon Dioxide Emissions in Power Generation

The major options that are alternatives to nuclear fission for decarbonizing the power sector are thermonuclear fusion, hydroelectric power, the "new" renewables (mainly wind and photovoltaic power), and decarbonized fossil fuels. Each is discussed in turn.

Thermonuclear Fusion

Fusion technology could provide electricity that is free of carbon dioxide emissions and might potentially be more proliferation-resistant than fission technology. Commercialization of fusion technology is not, however, expected before the middle of the century.[21] Even if nuclear fusion could be brought to commercial readiness in fifty years, its market prospects would be quite uncertain, because it would face much stiffer competition than nuclear fission faces today, as will be apparent below.

Hydroelectric Power

Hydroelectricity accounted for 19 percent of global electricity generation in 1997 (see table 1). The technical and economic potential for expanding hydroelectric power, a fully established renewable electric technology, is five and a half times and three times the 1997 generation rate, respectively;[22] in IS92a, the economic potential is fully exploited by 2100 (see table 1). Although large hydropower projects are coming under increasing attack on environmental grounds,[23] the future of climate-change mitigation is not very dependent on the future environmental constraints that hydropower faces. If hydropower expansion were limited to plants already under construction and the deficit were made up entirely by coal power, carbon dioxide emissions of carbon dioxide in 2100 would be only 5 percent more than in IS92a.

New Renewables

For new renewables, the focus here is on wind and photovoltaic power and on how use of appropriate electric storage technology could facilitate high levels of electric grid penetration by these intermittent renewables.

Coping with intermittency

The intermittency of wind and photovoltaic power is often viewed as a constraint on electric grid penetration. Gas turbines and/or combined cycles and/or hydroelectric power supplies can provide effective backup to these intermittent renewables. If those systems accounted for a substantial fraction of the capacity on the grid, it is possible to assure highly reliable power for grid systems at grid penetration levels by intermittent renewables up to 10–30 percent without new electric storage technology, despite the intermittency.[24] Without electric storage but with the grid backup capacity mainly in the form of nuclear or supercritical fossil steam plants, which cannot respond quickly to changing load conditions and for which idle capacity costs are high, much lower penetration by intermittent renewables is feasible. The implication is that nuclear and intermittent renewables represent competitive rather than complementary power technologies in grid applications.

Electric storage is needed to achieve higher levels of electric grid penetration. Intermittent renewables can provide either load-following or baseload outputs if coupled to suitable electric storage. Fortunately, technological breakthroughs are not needed because commercially available compressed air energy storage technology offers attractive costs.[25] Compressed air energy storage is well-matched to intermittent renewables because costs are much lower than for all other electric storage technologies with about a day of storage (see table 2),[26] as would typically be needed to make load-following or baseload power from intermittent renewables. Compressed air energy storage does require suitable geology: bedded or domed salt formations that can be solution-mined, mined spaces in hard rock, or porous media (aquifers or depleted natural gas fields). About 85 percent of the United States has one or more suitable geologies for compressed air energy storage.[27]

Wind power

The first modern grid-connected wind turbines were installed around 1980. By 1990 about 2 gigawatts of grid-connected wind capacity were operating worldwide. In 1999 installed wind capacity worldwide

Table 2. Capital Costs for Electrical Storage (1997 dollars)

	Component cost		*Total cost ($/kilowatt-hour)*	
Technology	*Discharge capacity ($/kilowatt)*	*Storage ($/kilowatt -hour)*	*2 hours*	*20 hours*
Compressed air				
Large (350 megawatts)	350	1	350	370
Small (50 megawatts)	450	2	450	490
Above ground (16 megawatts)	500	20	540	900
Conventional pumped hydro	900	10	920	1,100
Battery (10 megawatts)				
Lead acid	120	170	460	3,500
Advanced (target)	120	100	320	2,100
Flywheel target (100 megawatts)	150	300	750	6,200
Superconducting magnetic storage target (100 megawatts)	120	300	720	6,100
Supercapacitor target	120	3,600	7,300	72,000

Source: Based on a presentation by Robert B. Shainker (Electric Power Research Institute) to the Energy Research and Development Panel of the President's Committee of Advisors on Science and Technology, July 14, 1997, reproduced from Panel on International Cooperation in Energy Research, Development, Demonstration, and Deployment, President's Committee of Advisors on Science and Technology, *Powerful Partnerships: The Federal Energy Research & Development for the Challenges of the 21st Century* (Washington, D.C.: Office of Science and Technology Policy, Executive Office of the President, June 1999). Available at http://www.whitehouse.gov/WH/EOP/OSTP/html/ISTP/Home.html.

reached 14 gigawatts, and wind accounted for 0.2 percent of total global electricity generation. Since 1994 the global capacity of wind power for electric grid-connected applications has been growing at 27–33 percent per year.[28] For the United States, the estimated potential for wind power that can be exploited *practically* is 10,000 terawatt-hours per year (see table 3), about three times the total rate of U.S. electricity generation in 1997. At the global level, Grubb and Meyer estimated that the practically exploitable *onshore* potential is 53,000 terawatt-hours per year,[29] whereas a World Energy Council study estimated the potential to be 20,000 terawatt-hours per year.[30] The author's estimate is 43,000 terawatt-hours per year,[31] which is more than the electricity generation rate from fossil fuels plus nuclear power in 2100 under IS92a (see table 1).

Wind electricity costs have fallen sharply since the early 1980s. Currently, the unsubsidized cost is as low as 4 cents per kilowatt-hour

Table 3. Estimated Recoverable Wind Energy by Wind Power Class in the United States (terawatt-hours per year)

Region	*Class 4*[a]	*Class 5*[b]	Class 6[c]
United States[d]	9,200	690	490
Great Plains[d]	8,900	570	420

Source: See table notes for sources.

a. Average wind speeds greater than 5.6 meters per second and less than 6.0 meters per second at 10-meters height; the wind power density at 50 meters is 400–500 watts per square meter.

b. Average wind speeds greater than 6.0 meters per second and less than 6.4 meters per second at 10-meters height; the wind power density at 50 meters is 500–600 watts per square meter.

c. Average wind speeds greater than 6.4 meters per second and less than 7.0 meters per second at 10-meters height; the wind power density at 50 meters is 600–800 watts per square meter.

d. Wind power potentials are based on estimates of 415,000, 28,000, and 17,000 square kilometers of available U.S. land area in wind Classes 4, 5, and 6 (and higher), respectively, in D. L. Elliott, L. L. Wendell, and G. L. Gower, "An Assessment of the Available Windy Land Area and Wind Energy Potential in the Contiguous United States," Pacific Northwest Laboratories Report PNL-7789, Pacific Northwest Laboratory, Richland, Wash., 1991. These estimates are based on the assumption that 100 percent of wilderness and urban areas, 50 percent of forest lands, 30 percent of farm lands, and 10 percent of barren and range lands would be excluded from wind power development. The power generation potential is estimated for: (i) a hub height of 100 meters with year 2030 technology (1,412, 1,566, and 1,797 kilowatt-hours of net generation per year per square meter swept by the turbine rotor for wind power Classes 4, 5, and 6, respectively—see table 4) and (ii) wind turbine spacings of 5 rotor diameters across the wind and 10 rotor diameters downwind, so that the wind generation potential per unit of land area is (/200) times the power generated per unit area swept by the turbine rotor.

where there are high-quality (Class 6) winds and 5–6 cents per kilowatt-hour in regions with moderate-quality (Class 4) winds (see table 4)—significantly less than the cost of electricity from a new nuclear plant (see table 5). In light of expected continuing technological improvements, the costs of wind electricity are projected to be less than 4 cents per kilowatt-hour by 2020 in regions with moderate-quality winds (see table 4), a level that is comparable to the costs of electricity from coal integrated gasifier/combined cycle and natural gas combined cycle power plants (see table 5).

The cost of electricity for a large baseload wind farm/compressed air energy storage power system with a 90 percent capacity factor (that is, the average system output would be 90 percent of peak output) would be only about 0.7 cents per kilowatt-hour more than the cost of electricity from simple wind farms at wind power costs projected for 2020 in regions with moderate-quality winds (see table 4).[32] At 4.4 cents per kilowatt-hour, the cost of such power would be less than that for baseload power from either nuclear power or natural gas or coal power with decarbonization/carbon dioxide sequestration technologies (see table 5).

Most good wind resources are in regions remote from major markets for electricity.[33] However, large-scale exploitation of these remote resources would be feasible by constructing large (multi-gigawatt) wind farms coupled to compressed air energy storage units to produce baseload power that could be transmitted at acceptable transmission costs to markets thousands of kilometers from the generation sites via high-capacity (gigawatt-scale) transmission lines operated at high-capacity factors.[34]

One concern about wind power is its land-use intensity. Suppose that by 2100 wind power were developed to the extent of the author's estimate of practically exploitable wind power—some 43,000 terawatt-hours per year, or enough to provide almost two thirds of the total global electricity projected for 2100 under IS92a (see table 1). The land area that the wind farms would occupy would amount to about 1.4 percent of the land area of the inhabited continents. However, only 5–10 percent of the land on which wind turbines are deployed is actually used for the turbines and their foundations, access roads, electrical substations, and other infrastructure. Most of the rest

Table 4. Projected Electricity Cost for Wind Electricity

	1997		*2000*		*2005*		*2010*		*2020*		*2030*	
Wind farm capacity (megawatts)	25		37.5		50		50		50		50	
Installed capital cost (dollars per kilowatt)	1000		750		720		675		655		635	
Hub height (meters)	40		60		70		80		90		100	
Rotor diameter (meters)	38		46		55		55		55		55	
Wind power class[a]	4	6	4	6	4	6	4	6	4	6	4	6
Net kilowatt-hours of electricity produced annually per square meter swept by turbine rotor	1011	1372	1192	1596	1294	1671	1334	1711	1385	1765	1412	1797
Average capacity factor (percent)	26.2	35.5	30.2	40.4	35.1	45.3	36.2	46.4	37.6	47.9	38.3	48.7
Generation cost (cents per kilowatt-hour)												
Capital[b]	6.54	4.82	4.26	3.18	3.51	2.72	3.19	2.49	2.99	2.34	2.84	2.23
Operation and maintenance	1.18	0.87	0.93	0.70	0.57	0.44	0.57	0.44	0.57	0.45	0.57	0.45
Overhaul/replacement	0.21	0.15	0.16	0.12	0.08	0.06	0.10	0.08	0.07	0.05	0.06	0.05
Royalties to landowners	0.24	0.18	0.17	0.12	0.11	0.08	0.10	0.08	0.09	0.07	0.09	0.07
Total	8.17	6.02	5.52	4.12	4.27	3.31	3.96	3.09	3.71	2.91	3.56	2.80

Source: Based on Electric Power Research Institute and the Office of Utility Technologies, Energy Efficiency and Renewable Energy, U.S. Department of Energy, "Renewable Energy Technology Characterizations," EPRI TR-10949, Electric Power Research Institute, Palo Alto, Calif., December 1997.

a. Wind Classes 4 and 6 are defined in notes a and c of table 3.

b. For an annual capital charge rate of 15 percent.

is usable for other purposes such as growing crops and ranching. People in the remote areas where most wind resources are concentrated are likely to be less concerned than people in densely populated areas about the aesthetic impact of large wind farms, if wilderness areas are avoided. To the extent that remote wind farms would be concentrated in farming/ranching regions (as would be the case in the United States—see table 3), wind farm royalties would provide a major supplement to the income from farming/ranching: in the United States, income per acre would typically be greater than the net income from farming/ranching.[35]

Photovoltaic power

Although photovoltaic power costs have fallen substantially since the mid-1970s, at present the costs of electric generation for central-station photovoltaic power plants in areas with good insolation are several times the costs of wind generation in regions with good wind resources. However, photovoltaic power offers major advantages over wind power and other renewables in that small photovoltaic systems can be sited near users—for example, on residential building rooftops, commercial building facades, and roofs of parking garages—where the power is worth much more than in central-station power plants. Such decentralized generation is feasible because a photovoltaic system requires no system operators, causes no pollution, is not noisy, and has costs per kilowatt-hour that are not especially sensitive to scale. Already photovoltaics is the least-costly means of providing electricity to households with modest levels of demand at sites remote from electric grids, including rural households in developing countries.[36] Photovoltaic systems for grid-connected applications are not yet competitive, but installed costs for grid-connected residential rooftop applications have been falling sharply, from $17 per watt in 1984 to $9 in 1992 and $6 in 1996.[37] For thin-film residential photovoltaic systems, installed costs are expected to reach $3 per watt during the period 2005 to 2010. At this cost level, photovoltaic systems on rooftops of new houses are expected to be fully cost-effective for U.S. consumers in several regions where net metering is allowed[38] and where photovoltaic systems are financed with home mortgages.[39] The potential residential

Table 5. Electricity Generation Costs for Baseload Nuclear, Natural Gas Combined Cycle, and Coal Integrated Gasifier/Combined Cycle Power Plants

	Nuclear[a]	Natural gas combined cycle[b]		Coal integrated gasifier/combined cycle[b]	
Sequestered carbon dioxide	-	No	Yes	No	Yes
Efficiency (percent)	-	54.0	45.7	45.9	36.1
Emission rate (grams of carbon per kilowatt-hour)	0	90	15.7	184	23.9
Carbon dioxide disposal rate (grams of carbon per kilowatt-hour)	-	-	91	-	210
Capital cost ($ per kilowatt-electric)	1,700–3,100	416	907	1,114	1,514
Generation cost (cents per kilowatt-hour)					
Capital[c]	3.64–6.64	0.89	1.94	2.38	3.24
Operation and maintenance	1.40	0.24	0.52	0.64	0.86
Fuel	0.54	2.27[d]	2.68[d]	0.73[d]	0.93[d]
Carbon dioxide disposal (at $5 per tonne of carbon dioxide[e])	-	-	0.17	-	0.39
Total cost	5.58–8.58	3.40	5.31	3.75	5.42
Cost with $104 per tonne carbon tax[f]	5.58–8.58	4.34	5.47	5.67	5.67

Source: See table notes for sources.

a. A recent survey by country indicates that installed costs for new light-water reactors range from $1,700–3,100 per kilowatt (electric) (J. A. Paffenbarger and E. Bertel, "Results from the OECD Report on International Projects of Electricity Generating Costs," a paper presented at the IJPGC 98: International Joint Power Generation Conference and Exhibition, August 24–26, 1998; Paffenbarger is with the International Atomic Energy Agency, and Bertel is with the Nuclear Energy Agency). It is assumed that the costs of fuel and of operation and maintenance are, respectively, 0.54 and 1.14 cents per kilowatt-hour—average values for U.S. nuclear plants in 1998 (R. H. Williams, "Advanced Energy Supply Technologies," in *Energy and the Challenge of Sustainability*, World Energy Assessment [New York: Bureau for Development Policy, UN Development Programme, 2000, Chapter 8, pp. 273–329]).

b. The costs for 400-megawatt (electric) natural gas combined cycle and coal integrated gasifier/combined cycle power plants, based on General Electric gas turbine/steam turbine combined cycles with steam-cooled gas turbine blades (technology recently commercialized for natural gas combined cycles). The calculations presented are from Robert H. Williams, "Advanced Energy Supply Technologies," in *Energy and the Challenge of Sustainability*, World Energy Assessment (New York: Bureau for Development Policy, UN Development Programme, 2000) and based on analyses by Dale Simbeck, SFA Pacific: the engineering and contingencies are 10 percent of the process capital equipment costs; general facilities are 10 percent of the process capital equipment costs; and annual operation and maintenance costs are 4 percent of the capital cost. For the carbon dioxide sequestration cases, the costs for the capital, operation and maintenance, and energy penalties associated with compressing the separated carbon dioxide to 135 atmospheres to ready it for disposal are included; the costs of the carbon dioxide disposal are indicated separately.

c. Assuming the annual capital charge rate is 15 percent and the average plant capacity factor is 80 percent.

d. For natural gas and coal at $3.40 per gigajoule and $0.93 per gigajoule, respectively, the average prices projected for U.S. electric generators in 2020 by the U.S. Energy Information Administration (Energy Information Administration, "Annual Energy Outlook 2001: With Projections to 2020," DOE/EIA-0383 [2001], U.S. Department of Energy, Washington, D.C., December 2000).

e. The disposal cost is the cost for the pipeline carrying the carbon dioxide to the disposal site plus the cost of the disposal wells for injecting the carbon dioxide into a suitable geological formation plus the cost of associated surface facilities at the disposal site. For 400-megawatt natural gas combined cycle and coal integrated gasifier/combined cycle plants, a disposal cost of $5 per tonne of carbon dioxide is typical for the disposal of carbon dioxide at sites that are, respectively, 50 kilometers and 80 kilometers from the power plant at which the carbon dioxide is recovered.

f. With this carbon tax, the cost of coal integrated gasifier/combined cycle electricity is the same with and without carbon dioxide sequestration.

rooftop photovoltaic market in the United States under these conditions is estimated to be as much as 40 gigawatts for installed costs of $3 per watt.[40]

Major gains for photovoltaic performance, energy payback,[41] and cost are expected in the future. An assessment carried out jointly for the Electric Power Research Institute and the U.S. Department of Energy projects that, between 2005 and 2030, system efficiencies for thin-film photovoltaics will reach nearly 14 percent (up from 6–8 percent at present), and system costs for central-station applications will decline to less than $1 per watt.[42] Realizing such gains would enable photovoltaics to compete in many markets throughout the world.

Although the economic prospects for photovoltaics in central-station power markets are quite uncertain, it is likely that photovoltaics will become widely competitive in distributed grid-connected markets. It is not unrealistic to expect that they could ultimately be deployed at an average global per capita rate on the order of 1 kilowatt (electric), which would require a collector area of about 7 square meters per capita near users in areas characterized by average insolation.[43]

At the global level, electricity generation in 2100 could plausibly be based entirely on renewables[44] if, at that time:

1. total generation is at the rate projected under IS92a (68,000 terawatt-hours per year, see table 1);
2. wind power is generated at the rate estimated by the author for the practically achievable long-term potential (43,000 terawatt-hours per year);
3. hydropower and biomass power are produced at the rates projected for IS92a (9,000 terawatt-hours per year total, see table 1); and
4. the rest of the needed power (16,000 terawatt-hours per year) is provided by distributed photovoltaic power, which would require deployment of 0.8 kilowatts of photovoltaic capacity per capita in areas of average insolation.[45]

Decarbonized Fossil Fuels

Contrary to conventional wisdom, avoiding carbon dioxide emissions from fossil fuel use does not require abandoning fossil fuels. Feasible technologies and strategies exist that make it possible to extract most of the energy contained in fossil fuels while simultaneously recovering the carbon as carbon dioxide and preventing its release into the atmosphere. The issues involved concern the capacity, security, and cost of alternative carbon dioxide disposal options and the costs of separating the carbon dioxide from fossil energy systems and preparing it for disposal.

Outlook for carbon dioxide disposal

The options for carbon dioxide disposal include its storage in both deep ocean and porous geological media. Although ocean disposal has received the most attention, environmental concerns and other large uncertainties about its prospects have led to a shift in focus in recent years to geological (underground) storage—in depleted oil and natural gas fields, in uneconomic coal beds, and in deep saline aquifers.[46]

Carbon dioxide injection for enhanced recovery of hydrocarbons might become a profitable focus of initial efforts to sequester carbon dioxide.[47] There are about seventy-four enhanced oil recovery projects worldwide, most (sixty-six) in the United States. In 2000 oil production via enhanced oil recovery accounted for 4 percent of total U.S. oil production, a byproduct of which is the sequestration of 30 million tonnes of carbon dioxide annually. Most of the injected carbon dioxide comes from natural reservoirs of carbon dioxide,[48] but 5 million tonnes per year comes from anthropogenic waste sources.[49]

Another option is carbon dioxide injection for enhanced recovery of methane from beds of coal that cannot be mined.[50] Large amounts of methane are trapped in the pore spaces of many coals. Injection of carbon dioxide into the coals can sometimes lead to efficient methane recovery because typically carbon dioxide is twice as adsorbing on coal as is methane. Therefore carbon dioxide can efficiently displace the methane adsorbed on the coal:[51] As carbon dioxide moves through the reservoir, it displaces methane. The limited experience to date indicates that very little of the injected carbon dioxide shows up in the production wells until most of the methane has been produced,[52] so that prospects for permanent sequestration of the injected carbon dioxide appear to be good. Of course, carbon dioxide sequestration in the coal bed would prevent subsequent mining of the coal. However, large amounts of the coal in the ground cannot be mined.[53] Unlike enhanced oil recovery, enhanced recovery of coal-bed methane via carbon dioxide injection is not a commercially established technology, although one U.S. producer of coal-bed methane has been carrying out a commercial pilot project in the San Juan Basin since 1996.[54]

Sequestration of carbon dioxide in depleted oil and gas fields is generally thought to be a secure option if the original pressure of the reservoir pressure is not exceeded.[55] One estimate of the prospective global capacity for sequestration in such reservoirs[56] is 100 and 400 gigatonnes of carbon for oil and gas fields, respectively;[57] other estimates are as low as 40 and 90 gigatonnes for depleted oil and gas fields, respectively, plus 20 gigatonnes associated with enhanced oil recovery.[58] The range of estimates is wide because the suitability of the properties of the reservoirs for storage varies greatly and because oil and gas recovery may have altered the formations and affected the integrity of the reservoirs.

Deep saline aquifers are much more widely available than oil or gas fields. They are present in all sedimentary basins, with a total area of 70 million square kilometers (two thirds onshore and one third offshore), equivalent to more than half the 130-million-square-kilometer land area of the inhabited continents. Some sedimentary basins offer better prospects for carbon dioxide storage than others.[59] To achieve high storage densities, carbon dioxide should be stored at supercritical pressures,[60] which typically requires storage at depths greater than 800 meters.[61] The aquifers at such depths tend to be saline[62] and not effectively connected to the much shallower (depths of less than about 300 meters) freshwater aquifers people use.

Until a few years ago it was generally thought that closed aquifers with structural traps would be required for effective storage. The potential capacity for global sequestering in such traps is relatively limited—about 50 gigatonnes of carbon,[63] equivalent to less than ten years of global carbon dioxide production from burning fossil fuels at the current rate. However, a growing body of knowledge indicates that many large, regional-scale open aquifers with good top seals (very low permeability layers) can provide effective storage, if the carbon dioxide is injected sufficiently far from the boundaries of the aquifers that the carbon dioxide either never reaches the boundaries or, if it does, leaks at a rate that is sufficiently slow to be of little consequence to climate change.[64] The reason is the extraordinarily slow rates of carbon dioxide migration in such reservoirs (typically on the order of 1 centimeter per year), a phenomenon called "hydrodynamic trapping" of carbon dioxide.[65] With large aquifers, the carbon dioxide will eventually dissolve in the water ("dissolution trapping" of carbon dioxide), and with sandstone reservoirs containing certain clay minerals (but not carbonate reservoirs), the carbon dioxide will, after dissolving in the water, eventually precipitate out as a carbonate mineral ("mineral trapping" of carbon dioxide).[66]

If structural traps are not required for effective storage, the potential capacity of aquifers might be huge: Estimates range from 2,700 gigatonnes[67] to 13,000 gigatonnes of carbon.[68] By way of comparison, the estimated remaining resources of recoverable fossil fuels (excluding methane hydrates) contain 6,000–7,000 gigatonnes of carbon.[69]

There is a growing base of experience with carbon dioxide disposal in aquifers. Statoil is carrying out a large project that involves recovering the carbon dioxide contaminant in natural gas from the Sleipner Vest offshore natural gas field in Norway at a rate of 1 million tonnes of carbon dioxide per year, which it is injecting and sequestering in a nearby aquifer under the North Sea.[70] A project involving large aquifer disposal that is expected to commence within ten years will involve recovery of more than 100 million tonnes of carbon dioxide per year (equivalent to 0.5 percent of total global emissions from fossil fuel burning) from the Natuna natural gas field in the South China Sea (71 percent of the reservoir gas is carbon dioxide).[71]

There are also thirty-one smaller acid gas disposal projects in Alberta that involve recovery of carbon dioxide along with hydrogen sulfide from natural gas fields and injection of this mixture of gases (characterized by a wide range of relative concentrations) into both aquifers and depleted oil and gas fields for storage. These projects are using underground disposal of carbon dioxide/hydrogen sulfide because this strategy for responding to the regulations governing emissions of sulfur into the air is less costly than the alternative of recovering the hydrogen sulfide from the natural gas and converting it to elemental sulfur.[72]

The long experience with enhanced oil recovery and the growing body of experience with aquifer disposal, as well as extensive historical experience with underground gas storage, are contributing to a growing confidence within the scientific community that long-term sequestration of a significant fraction of global carbon dioxide production from human activities over the next several hundred years might be feasible.[73] Issues of public acceptability are, however, paramount. For most people, fuel decarbonization with carbon dioxide sequestration is an unfamiliar strategy for dealing with climate change. The scientific community has a major responsibility to inform the public debates on the issues relating to safety and environmental impact. Understanding both natural events[74] and the extensive historical experience with carbon dioxide injection for enhanced oil recovery and with underground gas storage[75] can help illuminate these issues. But more research, field testing, monitoring, and modeling are needed to narrow the uncertainties relating to carbon dioxide storage in geological media.

Outlook for carbon dioxide capture in power generation

Large central-station coal-fired power plants are likely to be early targets of efforts to recover the carbon dioxide associated with fossil fuel consumption and to prevent its release into the atmosphere by disposing of it in geological formations or otherwise.[76] For such systems, the cost of fuel decarbonization plus carbon dioxide disposal in a geological reservoir is made up of the costs of:

1. separating out a relatively pure stream of carbon dioxide from the fossil energy system;
2. compressing the carbon dioxide to a dense (supercritical) state for pipeline transport to the disposal site;
3. constructing the carbon dioxide pipelines and gaining associated rights of way; and
4. digging the wells for injecting the carbon dioxide into disposal reservoirs and constructing associated surface facilities.

Most decarbonization studies have focused on recovering carbon dioxide from flue gases of fossil fuel power plants. The cost of separating out carbon dioxide from flue gases dominates the costs of separation and disposal because the carbon dioxide concentration is low (~12 percent), as is its partial pressure (0.12 atmosphere). At such a low partial pressure, the recovery of carbon dioxide is typically achieved by making it react with amines (chemical solvents) to form a weakly bonded intermediate compound that is heated to recover the carbon dioxide stream and regenerate the original solvent. If carbon dioxide were recovered from flue gases of modern coal steam-electric plants, the cost to generate electricity would be 60 percent higher than without carbon dioxide recovery.[77]

A superior approach involves recovering carbon dioxide before combustion (so that it is undiluted by nitrogen from the air) in a coal integrated gasifier/combined cycle power plant. The coal integrated gasifier/combined cycle power plant is a commercial technology that makes it possible to convert coal to electricity at a much higher efficiency (about 46 percent) than with conventional coal steam-electric technology (35–36 percent for new plants and 33 percent for average U.S. plants in use). Air pollutant emission levels for this technology are comparable to those from natural gas combined cycles; moreover, costs of environmental damage associated with such emissions are about one-tenth of those for coal steam-electric plants equipped with the best available control technologies.[78] Even without taking credit for the environmental benefits offered by coal integrated gasifier/combined cycle power plant technology, the generation costs are approaching those for conventional coal steam-electric plants equipped with flue gas desulfurization.[79] Moreover, the costs of decarbonization/carbon dioxide sequestration are much less than those for coal steam-electric technology.

When coal integrated gasifier/combined cycle power plant technology is used, decarbonization/carbon dioxide sequestration begins with gasification (partial oxidation) of the coal in oxygen (obtained from an air separation plant) and steam at high pressure and temperature to produce synthesis gas (mostly carbon monoxide and hydrogen). The synthesis gas is cooled and scrubbed to remove pollutants (other than hydrogen sulfide, the compound formed from sulfur in the coal during gasification) and then passed to water-gas-shift reactors equipped with hydrogen sulfide-tolerant water-gas-shift catalysts. In these units, carbon monoxide is reacted with steam according to the water-gas-shift reaction (which reacts carbon monoxide with steam to form carbon dioxide and hydrogen). The shifted synthesis gas (made up mostly of hydrogen, carbon dioxide, and steam) is then cooled to near ambient temperatures, the hydrogen sulfide is removed using a glycol solvent and converted to elemental sulfur, the steam is condensed out, and the synthesis gas (at this point mostly hydrogen and carbon dioxide) is sent to a carbon dioxide separation unit. There the carbon dioxide at high concentration and partial pressure is recovered using a physical solvent[80] and made ready for disposal by drying and pressurization to 135 atmospheres. The hydrogen-rich synthesis gas is then burned in a combined cycle power plant. It is assumed here that the separated carbon dioxide is transported some 80 kilometers by pipeline to a site where it is injected via disposal wells into an appropriate storage medium (such as a deep saline aquifer).

The entire process can be accomplished using commercially available technologies and making relatively straightforward modifications to conventional coal integrated gasifier/combined cycle power plants from which carbon dioxide is vented. The estimated performance and electricity costs for both conventional and decarbonized coal integrated gasifier/combined cycle plants are summarized in table 5. As the table shows, an integrated gasifier/combined cycle plant located about 80 kilometers from the carbon dioxide disposal site and whose carbon dioxide emissions per kilowatt-hour are 87 percent less than those for a conventional integrated gasifier/combined cycle plant would have:

1. a capital cost that is 36 percent higher than that of a conventional integrated gasifier/combined cycle plant;
2. a conversion efficiency that is 10 percentage points less (but still higher than the 33 percent average efficiency for U.S. coal steam-electric plants); and
3. a generation cost that is about 45 percent higher.

A carbon tax somewhat greater than $100 per tonne of carbon would be needed to motivate decarbonization/carbon dioxide sequestration for integrated gasifier/combined cycle technology. Even with such a carbon tax, however, the cost of electricity from a decarbonized coal integrated gasifier/combined cycle plant would be less than that of a new nuclear plant except in those parts of the world where capital costs are at the very lowest end of the range of costs for new plants (see table 5).

One reason it is difficult for nuclear power to compete is the high operation and maintenance cost of existing light-water reactors (see table 5). Plausibly, new reactor technologies that have a high degree of inherent safety would have lower operation and maintenance costs. One such technology that has been commercialized recently is the 1,300-megawatt-electric advanced boiling water reactor developed by General Electric/Toshiba/Hitachi. One of these evolutionary light-water reactors is in operation in Japan, and the technology has been granted design certification in the United States. For the advanced boiling water reactor to be competitive with the decarbonized coal integrated/gasifier combined cycle power plant, its operation and maintenance costs would have to be about 30 percent less than the 1998 average value in the United States.[81] However, measures likely to be enacted to deal with security risks in light of growing concerns about nuclear terrorism will make it difficult to realize low operation and maintenance costs.

Still, the prospects are good that, wherever substantial wind resources are available, baseload wind/compressed air energy storage power will become less costly than decarbonized coal integrated gasifier/combined cycle power and nuclear power and thus would be preferred in a world seeking to constrain greenhouse gas emissions. As discussed below, the most promising opportunities for decarbonized fossil fuels are in the markets for fuels used directly.

Deep Reductions of Carbon Dioxide Emissions with Fuels Used Directly?

Even if the power sector were to be completely decarbonized by 2100 through the deployment of some mix of renewable, nuclear, and decarbonized fossil fuel technologies, carbon dioxide emissions from fuels used directly (other than for power generation) in 2100 in the IS92a scenario would still amount to 15 gigatonnes of carbon per year—several times higher than the levels required to keep the atmospheric concentration of carbon dioxide in the range of 450–550 parts per million. Reducing total global carbon dioxide emissions by 2100 to about 3 gigatonnes of carbon per year would be required to stabilize the atmospheric concentration of carbon dioxide at 450 parts per million. That level can be achieved only if the production and use of most fuels used directly at the levels indicated for 2100 in the IS92a scenario are characterized by near-zero net carbon dioxide emissions.

The main options for realizing zero or near-zero emissions from fuels used directly are:

1. hydrogen produced from fossil fuels with sequestration of the carbon dioxide coproduct;
2. hydrogen produced electrolytically from noncarbon (renewable or nuclear) power sources;
3. hydrogen produced by thermochemical splitting of water using heat generated from a noncarbon source (for example, a high-temperature gas-cooled reactor or a solar furnace); and
4. biomass fuels.

Each option is discussed in turn, following an introductory discussion of the prospects for introducing hydrogen as a major energy carrier.

Hydrogen as an Energy Carrier

In combustion applications, hydrogen can be burned as a fuel for cooking, for providing low-temperature heat (such as for space and heating of water), and for providing high-temperature heat for industrial processes.[82] The only air pollutants arising from hydrogen combustion are oxides of nitrogen, which can be controlled to very low levels by lean-burn combustion, steam or water injection, or catalytic combustion techniques.

Perhaps the most important market opportunity for hydrogen would be in fuel cells, because hydrogen is the natural fuel for that use. In fuel cell applications, even emissions of oxides of nitrogen would be negligible, so that hydrogen fuel cells would be characterized by zero emissions of both air pollutants and greenhouse gases. The prospects are good that during the next decade or so fuel cells will be commercialized both for stationary power and combined heat and power applications and for mobile applications. Fuel cell buses will soon be commercialized in various countries, and an intense race is underway among all the world's major automakers to commercialize fuel cell cars. Hydrogen fuel cell cars would be less costly to own and operate than fuel cell cars provided with gasoline or methanol fuel that is converted onboard to a hydrogen-rich fuel that the fuel cell can use. Breakthroughs are not needed in hydrogen storage before hydrogen fuel cell cars can be commercialized, because fuel-efficient fuel cell cars can be redesigned to accommodate commercially available compressed gaseous hydrogen storage. The prospects for successfully launching hydrogen fuel cell cars in the market during the second decade of this century are also good.[83]

Hydrogen from Fossil Fuels

Although hydrogen is not yet used as an energy carrier, it is widely manufactured for use in the chemical processing and petroleum refining industries. In the United States, about 1 percent of primary energy use (mostly natural gas) is committed to the manufacture of hydrogen. The process of making hydrogen from a fossil fuel begins with the conversion of the fossil fuel feedstock into synthesis gas. At present most hydrogen is manufactured from natural gas by steam reforming.[84] As noted, synthesis gas can be made from coal via gasification (partial oxidation). The manufacture of hydrogen from coal using commercial technology is essentially the same as the process described above for decarbonization of coal for integrated gasifier/combined cycle power applications, except that at the final stage following the removal of the carbon dioxide, the hydrogen-rich gas would be purified (up to 99.999 percent purity) in a pressure swing adsorption unit instead of being burned in a combined cycle power plant.

The process is more capital-intensive for coal than for natural gas, but coal is typically the less costly feedstock. For example, in the United States the average price paid by electricity generators for natural gas in 1997 was 2.1 times the price paid for coal, and this ratio is projected to increase to 3.7 times by 2020 (see note d, table 5), which is perhaps as soon as hydrogen could be used as an energy carrier in significant markets. At the U.S. coal and natural gas prices projected for 2020, hydrogen from coal is likely to be less costly than hydrogen from natural gas. It is estimated that, with near-commercial technology, hydrogen could be produced from coal with sequestration of the carbon dioxide coproduct at a cost of $6.4 per gigajoule and an efficiency of 72 percent, assuming coal was priced at $0.93 per gigajoule (see note d, table 5) and a $100 per tonne carbon tax.[85] Advanced technologies based on the use of inorganic membranes for separating hydrogen from carbon dioxide and other gases might lead to lower costs for making hydrogen from coal.[86]

In terms of energy content, this plant-gate cost of hydrogen is equivalent to a gasoline price of $0.92 per gallon, not much higher than the $0.88 per gallon average U.S. refinery-gate gasoline price in 2000. If this hydrogen were used by fuel cell cars, the cost of the hydrogen distribution and refueling system would increase the cost seen at the pump by consumers to about $2 per gallon of gasoline equivalent[87]—much higher than the average gasoline pump price in 2000 of $1.14 per gallon (excluding retail fuel prices). However, because fuel cell cars are expected to be two to three times as fuel efficient as today's cars with gasoline internal combustion engines, the cost of fuel per mile of driving would be less.[88]

Large-scale geological sequestration of carbon dioxide would be required if the coal-based hydrogen option were to be widely pursued in a world that is limiting greenhouse gas emissions. If in 2100 hydrogen derived from coal replaced all fossil fuels used directly under IS92a except those used to make iron and steel (assumed to be coal) and to run jet airplanes (assumed to be oil),[89] the amount of hydrogen required in 2100 would be 452 exajoules per year, and the carbon dioxide sequestration rate would be about 16 gigatonnes of carbon per year—2.6 times the total rate of carbon dioxide emissions from fossil fuel burning in 1997 (see table 1).

Consider how the global energy system might evolve to such an intensive level of sequestration. Suppose that by 2010 the initial carbon dioxide sequestration projects associated with coal conversion involve a rate of sequestration of 8.5 million tonnes of carbon dioxide per year (eight and a half times the rate of sequestration for the ongoing Sleipner project in the North Sea). If there were a public policy in place that aimed to expand this activity at an accelerated rate of 35 percent per year until 2025,[90] the incremental annual capacity for sequestration added in 2025 would be 770 million tonnes of carbon dioxide per year (210 million tonnes of carbon per year). If, after this rapid ramp-up, sequestration were continued at a linear rate of expansion of 210 million tonnes of carbon per year, each year from 2025 to 2100, the target rate for 2100 could be realized. Under this scenario the total amount of carbon dioxide sequestered in this century would be about 600 gigatonnes of carbon—a large amount, but one that is significantly less than even the pessimistic estimates of the global capacity for secure geological sequestration. To be sure, undertaking such a global activity will require a much better understanding of carbon dioxide storage and the risks involved than is available at present.

Electrolytic Hydrogen

An alternative to making hydrogen from fossil fuels with carbon dioxide sequestration is to make it by breaking apart water molecules using electricity from either nuclear or renewable electric supply sources. Consider, first, making hydrogen electrolytically from wind power at the cost projected for Class 6 winds in 2020—some 2.9 cents per kilowatt-hour (see table 4). A large wind farm providing power at this price coupled to a compressed air energy storage unit could provide baseload electricity at a relatively modest incremental cost. Such baseload electricity from a remote wind farm in turn could be transported via a high-voltage transmission line at a low incremental cost to an urban center, where it might be used to make hydrogen electrolytically for use in transport. It is estimated that the cost of such baseload electricity delivered 300 kilometers to a "city gate" would be about 4.0 cents per kilowatt-hour.[91] Moreover, by 2020 it is plausible that advanced electrolytic conversion equipment could become widely available and would result in a cost for hydrogen derived from wind power that is 25 percent less than with current electrolytic technology at an electricity price of 4.0 cents per kilowatt-hour. Still, the cost of the electrolytic hydrogen produced with advanced electrolytic technology would be two and a half times more costly than hydrogen produced from coal with carbon dioxide sequestration using near-commercial technology.[92]

Is it plausible that, with advanced technology, electrolytic hydrogen might one day emerge the economic winner? Unfortunately, the answer is no. Even assuming advanced electrolysis technologies, the economics of the production of electrolytic hydrogen from any carbon-free electricity source would always be unattractive unless there were "fatal flaws" associated with the option of sequestering the carbon dioxide byproduct of making hydrogen from fossil fuels. Consider the electricity price required for break-even. Assuming baseload (90 percent capacity factor) electricity, the electricity price for a carbon-free source would have to be 0.9 cents per kilowatt-hour in order for electrolytic hydrogen based on advanced electrolytic technology to be competitive with hydrogen produced from coal using near-commercial technology with sequestration of the separated carbon dioxide.[93]

It is unlikely that electricity prices below 1.0 cent per kilowatt-hour will be routinely achievable with either nuclear or renewable electric technologies, at least over the next several decades. Off-peak hydroelectric power prices are typically this low or lower, however, and eventually variable nuclear costs might plausibly become low enough to put nuclear off-peak prices in the targeted range as well. However, off-peak pricing strategies would be appropriate only where hydrogen production is a minor activity relative to electricity generation, so that power generation could shoulder the capital and other fixed charges. In a greenhouse gas emissions–constrained world, however, hydrogen is likely to be required in the late twenty-first century at levels far in excess of the level of power generation,[94] so that the fixed charges must be allocated to the production of hydrogen. Thus, electrolytic hydrogen generated via either nuclear or renewable electric sources is not economically promising, and neither would be considered a major energy option in this century unless presently unforeseen flaws emerge in the carbon dioxide sequestration option.

Thermochemical Routes to Hydrogen Production from Water

An alternative way to split the water molecule to produce hydrogen is by applying heat. Using this process directly requires temperatures of the order of 4,000°C—a level that cannot be accomplished at present because there are no materials that can contain the reactions. Over the years, however, various multiple-step chemical processes have been proposed for making hydrogen from water thermochemically at lower temperatures, using either nuclear heat (for example, heat that can be provided by a high-temperature gas-cooled reactor)[95] or high-temperature solar heat that can be provided with collectors that concentrate sunlight.[96] In contrast to electrolytic processes, thermochemical processes for hydrogen manufacture are far from being commercially available. Here, as in the electrolytic case, the prospects are bleak that thermochemical conversion would *ever* become economically competitive with fossil fuel-derived hydrogen with carbon dioxide sequestration.

Such processes tend to be quite capital-intensive and have overall thermal efficiencies for converting heat into hydrogen that are typically less than 50 percent.[97] The UT-3 process,[98] an option that has been investigated extensively in Japan, illustrates the challenge facing these technologies. For the most promising configuration of a system to produce hydrogen using the UT-3 process, a recent study estimated that the overall efficiency of converting nuclear heat into hydrogen would be 45 percent (higher heating value basis).[99] For this technology the estimated cost of producing hydrogen is $37.5 per gigajoule (see table 6)—about six times the $6.4 per gigajoule cost of making hydrogen from coal using near-commercial technology with carbon dioxide sequestration.[100]

One reason for the high cost is that nuclear heat is expensive, accounting for nearly 60 percent of the total production cost. Even if this heat were "free," the cost of hydrogen would still be $15.5 per gigajoule, some 2.4 times the cost of hydrogen from coal using near-commercial technology with carbon dioxide sequestration.

The prospects are not bright that more promising thermochemical cycles will be discovered and developed in the future. Suppose that some advanced process is discovered that is so cheap that costs other than for nuclear heat are essentially zero. Y. Tadokoro and his colleagues estimate that for the UT-3 process high-temperature gas-cooled reactor heat would cost $9 per gigajoule[101] (see table 6). Assuming this cost for heat and a free conversion technology, the pro-

Table 6. Estimated Costs for Thermochemical Production of Hydrogen from Water Using the UT-3 Process and Nuclear High-Temperature Gas-Cooled Reactor Heat ($/gigajoule, higher heating value basis)

Capital	12.53
Maintenance	2.31
Labor	0.42
Utilities	2.78
High-temperature gas-cooled reactor heat[a]	heat cost/0.449
Credit for byproduct oxygen	− 3.99
Subtotal	14.05 + heat cost/0.449
Overhead	0.1*(14.05 + heat cost/0.449)
Total production cost	15.46 + 2.45*(heat cost)
Total production cost for heat costing $9.0 per gigajoule[b]	$37.5

Source: Based on Y. Tadokoro, T. Kajiyama, T. Yamaguchi, N. Sakai, H. Kameyama, and K. Yoshida, "Technical Evaluation of UT-3 Thermochemical Hydrogen Production Process for an Industrial Scale Plant," *International Journal of Hydrogen Energy* 22 (1)(1997): 49–56.

Note: The figures are for a system producing, from water, 20,000 normal cubic meters of hydrogen per hour using heat from a high-temperature gas-cooled reactor via the UT-3 process under development in Japan. To facilitate a comparison of the costs presented in Tadokoro et al., however, it is assumed here that the capital charge rate is 15 percent per year (as for other hydrogen production technologies discussed in this paper) and that the system operates at an average annual capacity factor of 80 percent (so that the annual rate of producing hydrogen is 1.787 million gigajoules per year). Further, only the most economically attractive option presented in Tadokoro et al.—that using a membrane to separate the hydrogen from other gases at a high rate of hydrogen recovery—is considered here. The results presented here in dollars are based on an exchange rate of 101 yen per dollar, the average exchange rate for 1995–96.

a. The contribution of high-temperature gas-cooled reactor heat to direct cost for hydrogen manufacture is the cost of the heat (in dollars per gigajoule) divided by the conversion efficiency (0.449).

b. Tadokoro et al. estimate that heat from a high-temperature gas-cooled reactor is available at 3.8 yen per 1,000 kilocalories, which equals $9 per gigajoule.

duction cost for hydrogen would be $22 per gigajoule, which is 3.4 times the cost of hydrogen from coal with carbon dioxide sequestration based on near-commercial technology.

The cost of heat from future high-temperature gas-cooled reactors might be substantially less, as would be the case if recent very optimistic projections of capital and generation costs for the pebble-bed modular reactor could be realized. A crude estimate is that the nuclear heat so generated might cost as little as $2.6 per gigajoule.[102] Even at this optimistic heat cost level, however, the contribution of just the nuclear heat to the cost of thermochemical hydrogen would be $6.4 per gigajoule, so that all other components of the cost presented in table 6 would have to be reduced to zero to enable thermochemical hydrogen to compete.

In light of its poor economic prospects, thermochemical hydrogen would not be considered seriously unless geological sequestration of carbon dioxide associated with fossil energy–derived hydrogen proves unworkable for reasons than cannot presently be identified, or when the limits on the capacity for geological storage for carbon dioxide are approached in the longer term (more than a century into the future). Moreover, even if those conditions could be satisfied, nuclear hydrogen would have to compete with both electrolytic hydrogen derived from photovoltaic or wind electricity sources and with hydrogen derived thermochemically using high-temperature solar thermal processes.

Biomass Fuels

To the extent that carbon-based fuels can be produced from biomass grown on a sustainable basis, hydrogen would not be needed to provide fuels for direct use in a world that limits the emissions of greenhouse gases. The reason is that the growing and use of biomass do not lead to a net buildup of carbon dioxide in the atmosphere: The carbon dioxide released in combustion is balanced by the carbon dioxide extracted from the atmosphere during photosynthesis. Biomass-derived energy can be provided from residues from the production of agricultural and forest products and from biomass grown on plantations dedicated to the production of biomass for energy.

Biomass can be converted to fuels used directly for transportation and other applications by various routes, including biological processes to produce ethanol (for example, from woody biomass via enzymatic hydrolysis, the main focus of the Department of Energy's biofuels development effort), and the synthesis gas route, used to produce fuels such as methanol, synthetic middle distillates, and dimethyl ether.

The Second Assessment Report of the Intergovernmental Panel on Climate Change[103] assessed the long-term potential of obtaining biomass from various residue sources. It concluded that by the end of this century the primary energy potentially obtainable from residues could amount to up to about 90 exajoules per year, slightly more than the global production of natural gas at present.

The growing of biomass on plantations dedicated to energy could increase biomass energy supplies beyond what residues could provide. Bioenergy production is, however, very land use–intensive, requiring about 4 million square kilometers (3 percent of the land of the inhabited continents[104]) for each 100 exajoules per year of primary biomass grown[105] to provide additional energy supplies. A detailed review of biomass energy options[106] carried out for the World Energy Assessment[107] concluded that worldwide some 7 million to 14 million square kilometers of land (5–10 percent of the land area of the inhabited continents) is potentially available for producing biomass for energy purposes. That land is made up of excess agricultural, degraded, and unproductive lands and does not pose major conflicts with the use of land for food production. Nevertheless, large-scale biomass growing for energy in dedicated plantations is likely to be a contentious activity because of its land-use intensity. Some would argue that it is preferable to encourage the conversion of excess agricultural, degraded, and unproductive lands to wildlife habitat rather than the cultivation of biomass for energy, even considering the climate change mitigation and other benefits that biomass plantations could provide.

The analysis of bioenergy carried out for the World Energy Assessment concluded that the practical global potential for biomass production for energy (residues plus plantation biomass) over the long term is 100–300 exajoules per year.[108] This forecast suggests that the biomass option offers little if any potential to improve upon IS92a in terms of greenhouse gas emissions from energy over the longer term, because that scenario already involves the use of 205 exajoules of biomass primary energy in 2100 (compared with 865 exajoules from fossil fuels, see table 1).

Conclusion

Effectively addressing the major energy challenges will require radical technological change. There is a need for energy technologies that are affordable and that offer zero or near-zero emissions of both air pollutants and greenhouse gases. Moreover, decades of rapid growth will be needed in the deployment of new technologies that offer significant promise to address these challenges.

Although nuclear power offers the potential for zero emissions for the power sector, several concerns must be dealt with effectively in order to exploit this potential. The prospects for addressing concerns about reactor safety, technical issues associated with disposal of radioactive waste, and perhaps also concerns about costs are reasonably good. Still, gaining public acceptance of waste disposal plans is a major challenge. Moreover, in the longer term the nuclear weapon connection to nuclear power would move to center stage among concerns about nuclear power if it were developed to the high levels needed to make a dent in addressing the climate change challenge. The author is not optimistic about the prospects for achieving adequate proliferation resistance via technical fixes in a world with ten to twenty times as much nuclear capacity as it has at present (the increase that would be required if nuclear energy were to play a significant role in climate change mitigation). Concentrating nuclear technologies in large nuclear parks maintained under tight security and international control would, however, probably be an effective response to the proliferation concerns in a nuclear-intensive energy future.

Nuclear power also faces strong competition from both new renewable electric technologies (mainly wind and photovoltaic) and decarbonized fossil fuel/carbon dioxide sequestration technologies.

For renewables, costs have fallen sharply, and the prospects are good that over the next ten to twenty years wind and photovoltaic power will become widely competitive in central-station and distributed grid-connected power markets, respectively. No new technological developments are needed to deal effectively with the issue of intermittency. The overall land requirements would be quite modest for both wind and photovoltaic technologies, even in future energy scenarios in which wind and photovoltaics come to account for most global electricity generation.

Fossil energy decarbonization/carbon dioxide sequestration has emerged as a major new competitor in the race to zero emissions for both power generation and markets that use fuels directly. There is growing confidence in the scientific community that up to several thousand gigatonnes of carbon in the form of carbon dioxide can be stored securely in deep geological reservoirs, so that it might be possible to use hundreds of years of fossil fuel supplies with very little release of carbon dioxide into the atmosphere. To be sure, there are uncertainties regarding the environmental impact of geological disposal of carbon dioxide on a large scale, but the uncertainties should be greatly reduced over the course of the next decade.

With commercially available technology, coal power plants could be built that have near-zero emissions of both greenhouse gases and air pollutants and that would be cost-competitive with nuclear power.

The largest challenge facing climate change mitigation is posed not by the power sector, but by the transportation and other sectors that use fuels directly. These sectors account for about two thirds of the global emissions of carbon dioxide at present and probably an even larger fraction in the future under business-as-usual conditions.

The least-costly supply option for achieving deep reductions in carbon dioxide emissions for fuels used directly is via hydrogen production from low-cost and abundant fossil fuel feedstocks (for example, coal), with sequestration of the separated carbon dioxide. Hydrogen produced electrolytically using electricity from nuclear or intermittent renewable power sources would be far more costly, even considering advanced technologies. The same is true of hydrogen that might be produced from water via thermochemical cycles driven by nuclear heat or heat derived from high-temperature solar concentrating collectors. Such options would be considered seriously only if it turned out that carbon dioxide sequestration could not be carried out on large scales. Biofuels will be important, but because of land use constraints, their contributions in the long term are not likely to be much higher than the levels projected for 2100 under IS92a.

To sum up, there are plausible combinations of energy supply technologies that would make it possible to address all the major challenges posed by conventional energy. Addressing the challenges effectively in this century would, however, require extraordinarily rapid deployment rates over decades. It is unlikely that such rates of deployment can be realized under free energy market conditions. Therefore, public policies are called for that set goals for tackling the challenges, support research and development on promising options that might address the challenges, create market-launching incentives for radical technologies that offer great promise in realizing the goals, and foster competitive market conditions for widespread deployment after being launched in the market.

Establishing such policies and keeping them in place long enough to make a difference will require a high degree of support among the general public for the targeted technologies. Such policies can endure in democratic societies over the many decades needed to keep the expansion of capacity on track *only if the general public enthusiastically embraces, not just tolerates, the targeted technologies.*

Opinion polls indicate that, of the different clusters of technologies reviewed here, photovoltaics and wind power probably have the best chance of garnering broad public support. It is too soon to tell how the public will react to fuel decarbonization/carbon dioxide sequestration technologies and strategies, as they are still largely unfamiliar. The experience with public attitudes toward the disposal of radioactive waste is not encouraging. However, carbon dioxide is not radioactive and would not be harmful as long as the leakage rates can be kept low, and the prospects for keeping leakage rates low seem good. The prospects for getting broad public support would be much better if the focus in decarbonization/sequestration is on the technologies emphasized in this review that offer near-zero emissions of air pollutants as well as carbon dioxide and that are therefore as clean as renewable energy technologies, instead of pursuing the bandaid approach of removing carbon dioxide from the stack gases of fossil fuel power plants, a process that many people regard as environmentally unacceptable. It would also facilitate the building of broad public support if renewable energy advocates viewed this cluster of technologies as complementary rather than competitive in the quest for clean and climate-friendly technology.

It is difficult to imagine how public enthusiasm for nuclear power can be rekindled and sustained over many decades. Nuclear power has been around for a long time, during which the general public has developed strong opinions. A *sustainable* nuclear power renaissance is likely only if new nuclear technologies come into the market that most people judge to be decisively better than alternative energy technologies. Technologies such as the pebble-bed modular reactor appear (on paper) to be much better than current nuclear technologies, but renewable and emissions-free fossil energy technologies could become widely available in the same timeframe or earlier. Moreover, before trying to generate enthusiasm for nuclear power, the nuclear industry and interested governments would have to overcome the intense hostility to nuclear power that exists among various groups in many countries.

In the longer term, the real showstopper may be the nuclear weapons connection to nuclear power. Although this issue has not been on most people's radar screens until recently, the terrorist attacks of September 11, 2001, have thrust it into the limelight. This renewed concern would come into even sharper focus in a world that has much more nuclear power capacity than at present—perhaps stimulated by a diversion incident or two. The large international nuclear park option would greatly weaken the weapons link and might make most of the general public more comfortable. However, would national governments find this option acceptable? Giving up some degree of energy sovereignty seems especially difficult for those countries that have substantial nuclear power programs and whose decision to go nuclear was originally motivated by the perception that nuclear power offered a promising route to energy autarky.

Finally, even if the weapons link to nuclear power could be adequately weakened by the use of international nuclear parks, there is a risk that public policies and resources committed to resurrecting the nuclear option would weaken efforts to develop and commercialize non-nuclear technologies that could have far greater impact in climate-change mitigation.

A World with, or without, Nuclear Power?

Richard L. Garwin

The future of nuclear power and the future of the world are linked.[1] For some the linkage is positive, while for others it is strongly negative to the extent that they judge nuclear power and civilization to be incompatible.

The view presented here is that the United States could and should reduce its active nuclear weapon inventory to one thousand warheads in the next couple of years and should take measures to transfer the materials from other warheads and all weapon-usable materials to the civil inventory. Russia should do likewise. The ultimate future of nuclear weapons is in doubt, but questions regarding their prohibition will hardly change until the total number of nuclear weapons in the world falls to one thousand or so.

A Vast Expansion of Nuclear Power—Problems and Potential

The use of nuclear power involves at least six important considerations, each of which is discussed further below:

1. the proliferation of nuclear weapons;
2. waste disposal;
3. catastrophic accidents;
4. the radiation dose to the public that accompanies normal operations of nuclear power plants and the nuclear fuel cycle;
5. global warming; and
6. costs, including capital investment and the fuel cycle.

Nuclear Proliferation

A great negative of nuclear power to most people is its relation to nuclear weapons. Uranium enrichment plants, which fuel most of the world's reactors, can be used to enrich uranium to the range of 90 percent or greater, where it is an ideal material for nuclear weaponry. Reactors fueled with uranium create plutonium, the material of choice for nuclear weapons. Nuclear weapon primaries made with plutonium enable the burning of enriched or even normal uranium in thermonuclear fusion secondaries. With enough skill, they provide an essentially unlimited explosive yield.

Even nuclear reactors that burn almost 100 percent uranium-235 are potent sources of neutrons and in principle can be used to irradiate normal or depleted uranium to produce plutonium-239, and highly enriched uranium fuel is itself weapon-usable. The uranium-233 that can be employed as fuel in some reactors (and perhaps even the thorium/uranium-233 cycle used in some breeders or near-breeders) is an excellent material for nuclear weaponry. Uranium-233, however, becomes contaminated by an intense gamma-ray emitter a couple of years after its production, necessitating remote fabrication of weapons using this material.

No nuclear weapon state has used so-called reactor-grade plutonium to any great extent to make nuclear weapons. As is well known, reactor-grade plutonium extracted from normal spent fuel from the typical power reactor has more than 20 percent plutonium-240 and typically 60 percent or less plutonium-239. The plutonium-240 arises from the capture of a neutron on plutonium-239, as it remains in the reactor for the long-term, efficient burnup of uranium-235 content in the reactor fuel. Typically, light-water reactors use fuel enriched to

3.5–5.0 percent uranium-235, and the uranium-238 is a potent sink for neutrons, forming first uranium-239 and then, within days, neptunium-239 and plutonium-239.

The Soviet Union has operated graphite power reactors with very low enrichment, and the resultant plutonium is quite usable for nuclear weaponry. For that reason there is interest in terminating the operation of those three operating reactors in Russia. Furthermore, although Russia has made safety improvements since the 1986 Chernobyl accident, its graphite power-type reactors are regarded as substantially less safe than the light-water reactors with containment that are more common in the rest of the world.

Uranium reactors are a problem for clandestine weapons production during their normal operation, especially if the fuel is reprocessed. In that case the plutonium is separated and stored as plutonium oxide in 2-kilogram amounts in small, welded steel cans. Handling these cans poses no radiation hazard, so that in principle they could be stolen or diverted to make nuclear weapons. For a long time there was a myth that nuclear weapons could not be made with reactor-grade plutonium, but the 1994 report of the Committee on International Security and Arms Control of the National Academy of Sciences and a 1997 U.S. Department of Energy document dispelled that myth.[2] After careful analysis and declassification, in January 1997 the Department of Energy issued a public report on plutonium disposition that contained the following statement:

> At the lowest level of sophistication, a potential proliferating state or subnational group using designs and technologies no more sophisticated than those used in first generation nuclear weapons could build a nuclear weapon from reactor-grade plutonium that would have an assured, reliable yield of one or a few kilotons (and a probable yield significantly higher than that). At the other end of the spectrum, advanced nuclear weapons states . . . could produce weapons from reactor-grade plutonium having reliable explosive yields, weight and other characteristics generally comparable to those of weapons made from weapons-grade plutonium. . . . Proliferating states of intermediate sophistication could produce weapons with assured yields substantially higher than the kiloton range possible with a simple, first-generation nuclear device.[3]

The report states further:

> The disadvantage of reactor-grade plutonium is not so much in the effectiveness of the nuclear weapons that can be made from it as in the increased complexity in designing, fabricating and handling them. The possibility that either a state or sub-national group would choose to use reactor-grade plutonium, should sufficient stocks of weapons-grade plutonium not be readily available, cannot be discounted. In short, reactor-grade plutonium is weapons-usable, whether by unsophisticated proliferators or by advanced nuclear weapon states. Theft of separated plutonium, whether weapons-grade or reactor-grade, would pose a grave security risk.[4]

J. Carson Mark, who headed the Theoretical Division at Los Alamos National Laboratory for many years, wrote the most accessible early publication on this matter.[5] He stated that there were no difficulties of kind, and not much of degree, in handling reactor-grade plutonium as compared with weapon-grade plutonium. With the simplest type of implosion system a reactor-grade plutonium explosive would always have a yield exceeding 1 or 2 kilotons, and in some cases the yield would, by chance, be substantially more. Aside from the spontaneous neutron production of plutonium-240, the somewhat greater gamma radiation from reactor-grade plutonium would not be a problem for those making relatively few nuclear weapons. The greater heat output would have to be handled by design choices. The study of the Committee on International Security and Arms Control agreed with Mark's analysis and went further to state that nuclear weapons of somewhat greater sophistication could be made with substantially higher and reliable explosive yield,[6] a point on which the author is thoroughly convinced.

This unique and major problem with nuclear reactors already exists, and to an extent that is difficult to grasp. In round numbers, nuclear power supplies about 20 percent of the electrical energy used in the world and about 6 percent (the entire electrical energy sector accounts for one-third of the primary energy) of the world's primary energy consumption. Despite the belief that developed societies could manage with less energy than they now consume per capita, it is generally believed that approximately a doubling of energy consumption worldwide is desirable to achieve an acceptable standard of living. Although much of the world's consumption of energy is not sufficiently concentrated for the use of nuclear power, multiplying these numbers indicates that the existing world total of three hundred 1-gigawatt (electric) equivalent reactors would need to grow to some nine thousand (or a fraction thereof) if nuclear power were to supply all (or a fraction) of the world's future energy needs.

The amount of weapon-usable material produced thus far from the limited nuclear power industry and decades of operation is enormous. At present, in excess of 100 tons of plutonium have been separated in reprocessing operations, and more than 1,000 tons are still present in spent fuel. None of the spent fuel has been transferred to a mined geologic repository, whether as intact spent fuel or as reprocessed vitrified fission product waste. At the nominal 6 kilograms of plutonium per nuclear weapon, 100 tons of separated plutonium would suffice to make sixteen thousand nuclear weapons, while the 1,000 tons or more in spent fuel would make one hundred and sixty thousand weapons. Clearly the hazard is not that some terrorist group or emerging nuclear power will capture all of the spent fuel and build a force of thousands or hundreds of thousands of nuclear warheads. Rather, the hazard is that a few tens of kilograms could be stolen, purchased, or diverted for the production of a few or a few dozen nuclear weapons that could be used to hold even a large country hostage.

The solution is obviously to account for and guard the material not as bulk, but as discrete items. In an era of individual empowerment, where a gang of bandits can attack the Millennium Dome in London in an attempt to steal millions of dollars in diamonds from an exhibit, the world needs to rethink questions of security and reinforcement.

It is clear, however, that the hazard does not grow in proportion to the amount of plutonium. If it is all in one vast repository, access to the first tens or hundreds of kilograms poses the hazard, not the vast amount of excess plutonium beyond that. Therefore a key element in the protection of weapon-usable materials is a strong limitation on the number of storage sites.

There are other possibilities for massive transfer of nominally civilian nuclear materials and capabilities for the production of weapons.

What happens with a failed state that has a nuclear power system? Can its reactors be maintained safely? Will the world (under the International Atomic Energy Agency and UN Security Council), in the midst of chaos, move to guard the nuclear installations against the theft of weapon-usable material or sabotage? That scenario is not likely. What about a country in good standing with the Nuclear Non-Proliferation Treaty and the International Atomic Energy Agency that acquires a full-scale, off-the-shelf nuclear power industry and expertise and later abandons its membership in the treaty? What action is the world ready to take under those circumstances? More particularly, what is the United States willing to do?

All things considered, including the metric of "comparative risk," large-scale deployment of light-water reactors without reprocessing is a reasonable approach. However, the questions just raised that go beyond the normal constraints of the Nuclear Non-Proliferation Treaty deserve urgent international attention.

Waste Disposal

What is to be done with spent fuel, another parameter of nuclear power? Here a specific and emphatic recommendation is in order: Create competitive, commercial, mined geologic repositories for spent fuel and nuclear waste, and have the International Atomic Energy Agency certify them. The agency would also need to certify the acceptable forms of spent fuel and nuclear waste. In the era of globalization it is ridiculous to insist that Switzerland or Belgium or England each do the research and development and find within their limited territories a site for the geologic disposal of nuclear waste.

There are appropriate sites for competitive, commercial, mined geologic repositories in western Australia (the Pangea proposal), China, Russia, and the United States. At 25 tons of spent fuel per standard reactor year, the world's reactors produce some 8,000 tons of spent fuel per year. Were nuclear power to supply half the world's primary energy in the future, this amount would increase to some 120,000 tons of spent fuel per year—more each year than the ultimate capacity of the planned repository at Yucca Mountain. Where and how this amount of nuclear waste can be accommodated need much more research, and it is premature to opine either way. Ultimately, disposal under the deep seabed may be the solution, with continued surveillance to avoid poaching to obtain long-decayed spent fuel for its plutonium content.

It is clear that the nonproliferation regime needs attention. Some states claim that the nuclear weapon states, particularly the United States and Russia, have a commercial advantage because their nuclear facilities are not subject to inspection by the International Atomic Energy Agency. Further, to include them would require a doubling of the agency's budget, and the nuclear weapon states should have to pay the equivalent amount. However, what is that amount? The author has argued elsewhere[7] that nations that configure their nuclear industry in such a way as to ease the burden of inspection should pay less as a consequence. This recommendation has consequences. For instance, pyroprocessing (or other reprocessing in which a substantial amount of highly radioactive material is kept with the actinides to be recycled) might be more "proliferation-resistant" in that it would require less International Atomic Energy Agency resources to ensure that the material is not diverted or to respond to a potential diversion. Nevertheless, it should not be forgotten that this is a measure of the benefit of pyroprocessing to nonproliferation. It is not absolute, and there is as yet no indication that overall cost is reduced by processing before disposal, compared with the direct disposal of spent fuel.

Modular high-temperature gas-turbine graphite reactors, whether of the General Atomics design with prismatic fuel elements or the pebble-bed reactor design now pursued by ESKOM in South Africa, should also be welcomed—as long as they meet the other criteria defined above.

Ultimately the matter comes down to comparative risk. Any state or group dedicated to acquiring nuclear weapons can choose the traditional route of building an enrichment capability or production reactors that are not burdened with the high temperatures and pressures of systems that produce electrical power. In the modern era they can resort to buying or stealing weapon-usable material or nuclear weapons. It is senseless to pay an appreciable amount of the cost of the house to reinforce the back door if the front door is left open. These technological approaches to the acquisition of nuclear weapon–usable material become easier with the passage of time.

In contrast, there is always the possibility of buying a full-feature, off-the-shelf nuclear power industry and waiting until the need or opportunity arises to build nuclear weapons. The nuclear industries of several countries appear to have been motivated by this contingency.

Catastrophic Accidents

Next is the question of catastrophic reactor accidents. Both the Ford-MITRE study in 1977 and the American Physical Society study on the safety of light-water reactors[8] recognized that those reactors, or even graphite reactors, could not produce a nuclear explosion with the creation of enormous amounts of radioactive materials. However, they could suffer catastrophic accidents that would liberate much of the existing radioactivity produced over years of operation. Since a reactor in one day produces as much radioactivity as a 50-kiloton nuclear explosion, and the fuel in a reactor has typically been there for an average of two years, a typical nuclear reactor has in its core the long-lived radioisotopes from 30 megatons of fission.

According to the reports of the UN Special Committee on the Effects of Atomic Radiation, the 528 atmospheric nuclear tests will contribute three hundred thousand cancer deaths (based on the International Committee on Radiation Protection's figure of 0.04 cancer deaths per person-sievert). According to the same calculation, Chernobyl, having contributed some 600,000 person-sieverts to global exposure, will be responsible for some twenty-four thousand cancer deaths. Although in its 1993 report the UN Special Committee on the Effects of Atomic Radiation refers to global exposure of 600,000 person-sieverts, that figure is nowhere to be found in its 2000 report. That omission raises a fear that not only has history been rewritten, but also that primary data from which hazards or risks could be estimated have been eliminated.

The 1977 Ford-MITRE book estimated that a single reactor accident in which the core melted down and all of the content was liberated to the atmosphere might kill 10,000 people and render vast territories uninhabitable at current standards. The 1979 accident at

Three Mile Island was consistent with that judgment that a core meltdown was no less probable than the experience thus far achieved in nuclear reactor operations, in contrast to the Atomic Energy Commission's judgment of one core meltdown in a million reactor-years. The 1986 Chernobyl accident, which involved a reactor that did not have the containment that is standard on light-water reactors, had consequences similar to the maximum accident of the 1977 study. As with the Challenger space shuttle accident, in which the specific failure that actually occurred was ignored because it was "too safe to fail," it is clear that reactor accidents were too horrible to think about.

After Three Mile Island II, probabilistic risk assessment came into its own and resulted in probability charts that extend to large reactor accidents involving the deaths of ten thousand people. Ultimately, probabilistic risk assessment and the prevention of risk come down to the exchange value of lives for dollars. Half a century ago large construction projects expected one death per $1 million expended, a figure that now in the United States is perhaps only one death per $100 million. Recent papers on the value of a life spared (or rather of a premature death avoided) state that a figure like $10 million is more representative.

Reactors such as high-temperature modular gas-turbine reactors can help eliminate the risk of catastrophic accidents. One type is being developed in Russia by General Atomics Corporation and another, with encapsulated fuel spheres formed into pebbles in a pebble-bed reactor, in South Africa. Claims are made that this approach is substantially cheaper than normal light-water reactors. How much it is worth to avoid a fatality seems to depend strongly upon whose money is involved—yours or other people's. It is possible to avoid many of these conceptual problems as long as only comparative risk is considered, but when comparative risk has to be evaluated together with comparative cost, the dollar/life coefficient enters unavoidably. The Ford-MITRE study concluded that each reactor-year of operation would involve one to two deaths on average. Contributing factors would be the nuclear fuel cycle and normal exposure from reactor operations, as well as the very small probability of a catastrophic accident.

Radiation Dose

The author's recent review of probabilities found that mining and milling of uranium in many cases contribute substantially more risk than the estimate in the Ford-MITRE study.[9] Until very recently the reprocessing of spent fuel contributed far more than did the normal operations of reactors whose fuel is reprocessed.

The failure to present the risk from reactors honestly is a major concern. As mentioned, the 1993 report of the UN Special Committee on the Effects of Atomic Radiation[10] found a 600,000 person-sieverts exposure for the global population, a figure that does not seem to appear in the 2000 report.[11] At the same time, there is a substantial effort to replace the estimate of "0.04 cancer deaths per person-sievert" by zero deaths, simply because no specific cancer deaths can be "attributed" to such dispersed radiation exposure.[12]

Table 7 provides a comparison of the radiation hazard of electrical power production from nuclear plants and that from coal plants, with only the ionizing radiation taken into account and not the much larger hazard posed by emissions of sulfur oxide and heavy metals from coal into the atmosphere. It is clear that mining and milling are substantial contributors to the hazard of nuclear power, whether there is reprocessing or not, but in the not-so-distant past reprocessing has swamped the radiation exposure from normal operations of that same reactor.

Recent information from British Nuclear Fuels Ltd. shows that its reprocessing operation at Sellafield now contributes substantially less than the figure in the table, in large part because it does not release radiocarbon (carbon-14) into the atmosphere. For 1997, for instance, British Nuclear Fuels Ltd. reports a total ten-thousand-year dose to the world population of 12 person-sieverts per reprocessed gigawatt (electric) per year of fuel,[13] in contrast to the 217 person-sieverts found in the 1993 report of the UN Special Committee on the Effects of Atomic Radiation.[14] The difference between the nine deaths per reprocessed gigawatt (electric) per year (217 times 0.04) and the 0.5 deaths (12 times 0.04) is striking. Moreover, environmentally conscious mining and milling can reduce the contribution from those operations by a factor of one hundred below the average shown in the table. Since 100 person-sieverts per reactor year mean, according to

Table 7. Collective Effective Dose to the Public from Effluents of the Nuclear Fuel Cycle (dose commitment in person-sieverts per gigawatt [electric]-year of operation)

Source	*Once-through and recycle*[a]	Reprocessing[b]	Coal[c]
Local and regional component			
Mining	1.1	0.9	0.002
Reactor/power plant operations (atmospheric)	1.3	1.3	20.0
Subtotal	2.4	2.2	20.002
Solid waste and global component			
Mine and mill tailings (release over 10,000 years)	150	120	
Reactor operation, disposal of intermediate waste	0.5	0.5	
Reprocessing, solid-waste disposal	0	1.2	125[d]
Reprocessing, globally dispersed radionuclides (to 10,000 years)	0	217[e]	
Subtotal	150	339	125
Grand total	152	341	145
Cancer deaths/ gigawatt (electric)-year (at 0.04/ person-sievert)	6	14	6

Source: Author's calculations, using data from the UN Special Committee on the Effects of Atomic Radiation, "Sources and Effects of Ionizing Radiation," report to the General Assembly, with scientific annexes, United Nations, New York, 1993.

a. From table 53, p. 200.

b. From table 53 and from table 42 for local and regional and table 51 for globally dispersed. This column is per gigawatt (electric)-year of a nuclear plant using reprocessing and recycle.

c. From p. 56, para. 143, and p. 57, para. 151. The reprocessing and recycle approach is assumed to use 20 percent less uranium than the once-through option.

d. Buildings constructed with 5 percent of the ash from power production.

e. Recent experience from British Nuclear Fuels Ltd. for 1997 shows that when carbon-14 is captured this number falls to 12 person-sieverts per gigawatt (electric)-year of reprocessed fuel. British Nuclear Fuels Ltd., Director of Safety, Health, and Environment, "Annual Report on Discharges and Monitoring of the Environment," Risley, Warrington, Cheshire, WA3 6AS, United Kingdom, 1997.

the International Committee on Radiation Protection coefficient, four cancer deaths worldwide per year, it is worth taking measures to reduce this exposure, only a tiny fraction of which comes from operations per se.

Global Warming

The typical light-water reactor converts 30 percent of its heat energy into electrical energy. A modern coal-fired plant (or a plant fired by natural gas) converts 40–50 percent of the primary energy into electrical power. However, the carbon dioxide emitted into the atmosphere from a plant fired with fossil fuels remains there for forty to one hundred years, and each year thereafter the enhanced greenhouse effect supplies the earth with as much heat as combustion does in the year of operation. Therefore a nuclear power plant built and operated for forty years would contribute less heat to the earth in a single year's operation than would a fossil-fuel plant.

The comparison of risk can be carried out further by burdening each ton of coal consumed with a carbon tax of $100 per ton to avoid global warming. For a coal-fired power plant to provide 1 gigawatt (electric) requires the annual consumption of 3 million tons of coal, or a $300 million carbon tax, about equal to the total revenue of a 1-gigawatt (electric) nuclear plant.

Such estimates of the appropriate carbon tax are not capricious and ought to enter strongly into the decision whether to support and expand nuclear power. As Robert Williams convincingly shows, carbon sequestration can help convert thousands of gigatons of inexpensive coal into a benign and flexible fuel not only for stationary power plants but also to fuel a hydrogen economy.

Cost of Nuclear Power

The various categories of hazards and benefits above—proliferation or nonproliferation, reactor accidents, disposal of spent fuel, and global warming—are all hypothetical, probabilistic, or long delayed. What most affects decisions is the cost of nuclear power in comparison with the costs of other approaches to providing electrical energy or heat.

Since the Three Mile Island accident in 1979, not a single nuclear plant ordered before that time has been built in the United States. However, many have been built elsewhere in the world, at a cost of perhaps $2–4 billion per gigawatt (electric). The cost of nuclear power is largely the capital investment. The cost of fuel and operations is on the order of 10–20 percent of the annual cost component representing the initial investment. Nevertheless, conventional uranium reserves have appeared to be inadequate for a vast expansion of nuclear power using existing types of reactors. From the beginning of the nuclear era the prospect of breeding additional fissile material (plutonium-239) from the 99.3 percent of natural uranium that is uranium-238 has been a vision of plenty—in contrast to the 0.71 percent fissile uranium-235 content of that same uranium.

With regard to the fast-neutron breeder reactor, which has always been associated with the long-term future of nuclear power, Edward Teller said:

> I have listened to hundreds of analyses of what course a nuclear accident can take. Although I believe it is possible to analyze the immediate consequences of an accident, I do not believe it is possible to analyze and foresee the secondary consequences. In an accident involving a plutonium reactor, a couple of tons of plutonium can melt. I don't think anybody can foresee where one or two or five percent of this plutonium will find itself and how it will get mixed with some other material. A small fraction of the original charge can become a great hazard.[15]

In fact, the cost of expanding and continuing nuclear power may be far less than nuclear power technology enthusiasts have supposed. They have usually jumped to consider breeder reactors because of the "shortage" of uranium fuel. With proven reserves of some 3 million tons of natural uranium, and consumption of some 200 tons per year per 1-gigawatt (electric) reactor, this resource would last for only about 15,000 reactor-years, or 50 reactor-years at a consumption of 300 reactors equivalent, or a mere two years if the reactors are to supply half of the world's future total energy needs.

Of great interest are the terrestrial "reasonably assured resources" of uranium, which are likely to amount to 100 million to 300 million tons of uranium at a price of $350 per kilogram (in comparison with the current spot market price of $20–30 per kilogram).[16] Nobody of right mind would buy uranium at $350 per kilogram when the same material is available at $30 per kilogram, but it is of primary importance to note that at $350 per kilogram these high-cost terrestrial resources would still be cheaper than the cost of recycling fuel in a light-water reactor (perhaps at a cost of $700 per kilogram of natural uranium avoided) or of building a breeder reactor with a capital cost that might be double that of a light-water reactor.

Ultimately, the world may have safe, economical breeder reactors, but it need not rush to perfect them; there are centuries available to do so. The reason is that, in addition to the 200 million tons of terrestrial high-cost uranium, there are 4 billion tons of uranium in the oceans—enough for two thousand years of operations of ten thousand light-water reactors. Half of this seawater uranium could be harvested without a substantial increase in cost above that of harvesting the first seawater uranium in bulk. That cost might run from $100–1,000 per kilogram, which is still probably cheaper than recycle and breeders. Even at the higher figure the cost of fuel is still affordable for ordinary reactors, and negligible for breeders.

If all enrichment costs and tails fraction remain the same, buying 200 tons of uranium at $1,000 per kilogram to fuel a typical light-water reactor for a year would be $200 million—approximately double the cost of power from a light-water reactor. However, the additional cost per kilowatt would be some 2 cents per kilowatt-hour, which can easily be afforded in comparison with the 10 or 20 cents per kilowatt-hour charged to the consumer and the 40 or 70 cents recently experienced in California.

In principle, seawater uranium is available to any producer and would be an article of commerce. The estimates of $100 to $300 per kilogram come from French and Japanese groups.[17] A recent paper provides an estimate of $1,000 per kilogram.[18] More such analyses are needed. The following comments on the paper by T. Kato et al. should not be seen as an attack but rather as a catalyst for additional work. A sounder estimate, whether it supports a high or low cost for seawater uranium, is important to the evolution of nuclear power in the next half century.

Kato et al. consider as a unit a plant capable of extracting 200 tons of uranium per year from seawater—enough to supply fuel continuously for a single 1-gigawatt (electric) power reactor (at current tails fraction). At 100 yen per dollar, the investment cost for an ashore facility is $269 million, of which $16 million are for chelating resin to retain uranium and $253 million are for equipment; the costs of transport ships amount to $66 million; and ocean facilities account for $1,721 million, of which $1,045 million are for an ocean floating facility and only $82 million for the primary absorbent.

Although the Kato study is more detailed and perhaps more realistic in its costing than previous estimates, it has analyzed the wrong system. It looks at an ashore facility because the authors reject the environmental hazard of a 2,000-ton ship with a load consisting largely of 15 percent hydrochloric acid. Far more dangerous loads are carried every day over the seas. Therefore a cost and absolute risk analysis of what would be a much cheaper system, such as that sketched by Foos et al., with the processing aboard ship,[19] is clearly needed. This type of processing would enable uranium farming in vast ocean areas far from shore.

In his book, *The Mythical Man-Month*, Fred Brooks, the architect of the IBM-360 computer line of the 1960s, wisely counsels that people should "plan to discard the first one."[20] That is, the first large project (computer operating system, for instance) should not be put into production but should instead be a training exercise. The design should be analyzed, criticized, and used as a stepping-stone to a second version, which might be marketed. That is exactly what is needed in the design of seawater uranium farms. In particular, even if the decision were to use ashore facilities, it seems unnecessary and costly to have ocean floating facilities on the surface of the sea. Instead, the buoyed adsorbent structures should be no closer to the surface than 30 meters, with the vertical strings of adsorbent beds loaded vertically into the ship either for on-board processing or for transport to a structure in the neighborhood. Since even at a cost of $300 per kilogram of uranium the investment per reactor in a uranium farm would amount to about $700 million, it is worth planning substantial research and development and design refinement to reach a minimum cost. Specialized design of the lift system to bring the adsorbent bed strings aboard is one candidate. A longer run prospect is the use of a nonstandard ship with a small waterline area that would be largely immune to heavy seas and would improve the fraction of the time the ship could operate.

It must be recognized that an annual compensation rate of 10 million yen per person-year for Japanese workers will make uranium farming attractive for organizations with much lower labor costs. Japan now buys uranium of terrestrial origin, and it likely will buy seawater uranium as well, and be the richer for doing so.

Conclusions and Recommendations

Recommendations for the future of nuclear power are as follows:

1. Prepare authorized competitive, commercial, mined geologic repositories.
2. Reinforce and further increase support to the International Atomic Energy Agency and the UN Security Council to provide not only an accounting function but also a protective function to safeguard nuclear reactors and the nuclear fuel cycle, and internalize the costs.
3. Provide honest assessments of the probabilities of accidents and risks in evaluating nuclear power and reducing the hazards of accidents.
4. Recognize the benefits of nuclear power in comparison with the forty to one hundred times larger contribution that fossil fuel without sequestration of carbon dioxide makes to global warming per unit of electrical energy.
5. For governments, spend good money now to determine and reduce the cost of acquiring uranium from seawater as a way to guide decisions on nuclear power and energy.

Recommendations for fossil and renewable power are as follows:

1. Adopt a regulatory framework that encourages both distributed generation, including wind and solar energy and combined heat and power, and co-generation.
2. Plan for massive transport of electrical power from sources to consumers, with transmission contributing to load leveling, for example, through superconducting power transmission lines.
3. Take storage seriously, especially compressed-air energy storage.
4. Demonstrate options for carbon sequestration for the clean use of coal in the production of electrical power and hydrogen.

It is also essential to reduce nuclear weapons and provide negative and positive security assurances so that the remaining, much smaller, number of nuclear weapons can be seen as contributing to the security of all, and not to the security of a few at the expense of the many. To this end, guarantees of nuclear weapon use on the part of international coalitions and the United Nations itself should be considered.

Can the world do without nuclear power? Yes, it can with carbon sequestration until the supply of coal is largely exhausted. Can the world live with nuclear power? Yes, it can if the risks and benefits are honestly acknowledged, and if the organizational and financial resources are committed to ensure against catastrophic accidents and nuclear weapon proliferation from the nuclear power system.

In a word, the author's judgment is an emphatic and unequivocal "maybe."

PART II

Can Nuclear Power Be Made Proliferation-Resistant and Free of Long-Lived Wastes?

8

Attempts to Reduce the Proliferation Risks of Nuclear Power: Past and Current Initiatives

Marvin Miller

At the same time that Amory Lovins was posing the question, "Can we have nuclear power without proliferation?" in his book *Soft Energy Paths*,[1] a prominent nuclear scientist, Edward Teller, was wrestling with the same issue in a series of lectures on the energy problem delivered at the Technion in Israel. Commenting on opposition to nuclear power on the grounds that it would give both subnational groups and nations access to weapon-usable materials, Teller argued that increased security could handle the subnational threat, but then added:

> I wish I could be as optimistic and positive about the remaining objection: as nuclear reactors spread among nations their production will enable almost every country to acquire nuclear weapons. This statement, most unfortunately, is true. I believe that eventually nuclear proliferation is unavoidable unless we find better solutions to international problems than are now on the horizon.[2]

Teller was neither the first nor the best-known nuclear scientist who was concerned that nuclear power would facilitate the acquisition of nuclear weapons. Enrico Fermi expressed similar sentiments: "It is not certain that the public will accept an energy source that produces vast amounts of radioactivity as well as fissile material that might be used by terrorists."[3] There is the oft-quoted statement in the Acheson-Lilienthal Report of 1946 that stressed the inadequacy of international inspections to prevent proliferation:

> There is no prospect of security against atomic warfare in a system of international agreements to outlaw such weapons controlled only by a system which relies on inspections and similar police-like methods. The reasons supporting this conclusion are not merely technical, but primarily the insuperable political, social, and organizational problems involved in enforcing agreements between nations each free to develop atomic energy but only pledged not to use bombs. . . . So long as intrinsically dangerous activities [that is, production and use of weapon-usable materials such as plutonium and highly enriched uranium] may be carried out by nations, rivalries are inevitable and fears are engendered that place so great a pressure upon a system of international enforcement by police methods that no degree of ingenuity or technical competence could possibly hope to cope with them.[4]

In the euphoria generated by Atoms for Peace, most people reassured themselves that the type of safeguards system that the Acheson-Lilienthal Report had judged to be inadequate could nevertheless minimize the risks of proliferation. Proliferation might be a problem down the road, but it was difficult to stand in the way of a technology that would "make the deserts bloom" and would be "too cheap to meter." Both developing and developed countries were eager to avail themselves of the benefits of this new energy source. Publicly, they meant peaceful use, that is, for power, desalination, and production of special isotopes for medicine and agriculture. Thus nuclear technology flowed out of countries such as the United States and the Soviet Union, and foreign students and scientists flowed in, eager to learn the tricks of the nuclear trade.

It was clear, however, that some of the same technologies, materials, and manpower could be applied to making weapons, and that safeguards could not prevent their diversion to such use. The wake-up call on the linkage between the peaceful and military atom was the test by India in 1974. Much has been written about the Indian nuclear program.[5] One point is worth noting here: Homi Bhabha's strong stance that India would never accept "colonialism in the nuclear sphere," coupled with the eagerness of nuclear suppliers to sell their wares, provided India with the opportunity to produce plutonium and then use it in a "peaceful nuclear explosive."

Subsequently, both national and international initiatives focused on minimizing the risk that civilian nuclear activities could be used as a cover for a weapons program. Since the United States was the key player in these efforts, it is important to look at the reassessment of nonproliferation policy in this country that started at the end of the administration of President Gerald R. Ford and that the incoming administration of President Jimmy Carter pursued vigorously.

The Plutonium Economy: Problem or Solution?

The debate about the connection between nuclear power and nuclear weapons was particularly heated during the Carter administration. The concern of the Carterites stemmed from the 1974 test by India and the prospect that a rapid spread of nuclear power after the 1973 oil crisis would provide the rationale for the acquisition of the materials and technologies that the Acheson-Lilienthal Report had deemed dangerous: highly enriched uranium and plutonium, as well as the uranium enrichment and reprocessing technologies that can produce these materials from natural or low-enriched uranium and irradiated reactor fuel, respectively.

The counterargument—that international safeguards at enrichment and reprocessing plants would be able to detect and hence deter the production of significant quantities of weapon-usable materials—was met with skepticism that detection could occur in a timely manner, that is, before their use in weapons. It was also pointed out that safeguards can only be effective when applied, and states not party to the Nuclear Non-Proliferation Treaty were not legally bound to accept safeguards on their indigenous facilities. Moreover, the treaty states could legally withdraw from the treaty on three months' notice, and no sanctions were specified for violations of treaty commitments.

In sum, the only proliferation-resistant fuel cycles were those in which neither highly enriched uranium nor plutonium were used in separated form. Any reprocessing or enrichment should preferably take place under international or multinational control. The Carter administration sought to implement these views by both domestic legislation and international persuasion, with a focus on eliminating commercial use of plutonium. However, although the International Nuclear Fuel Cycle Evaluation program, organized at the behest of the Carter administration, supported the need to raise the level of consciousness about proliferation, in Western Europe and Japan there was strong resistance to delegitimizing their use of plutonium in the nuclear fuel cycle.

In brief, the Europeans and Japanese, and their supporters elsewhere, argued that:

1. nuclear power was essential, and there was not enough uranium to support a large nuclear enterprise if only once-through fuel cycles were utilized; and
2. the proliferation risks involved in plutonium use were exaggerated, since:
   - it was difficult to make reliable nuclear weapons using reactor-grade plutonium; and
   - once-through fuel cycles were not a panacea, since it was easy to extract plutonium from power reactor spent fuel in a "quick and dirty" reprocessing plant.

Both of these contentions were discussed at great length at the time, and the former continues to cause controversy despite the efforts of knowledgeable individuals such as the late J. Carson Mark,[6] Richard Garwin,[7] and John Kammerdiener[8] to shed light on the subject. Part of the problem is that key aspects of weapon design are still classified, including how the obstacles involved in using reactor-grade plutonium need to and can be overcome. Another consideration is that the stakes are high, which tempts some people to pay "selective attention" to the facts. In lieu of an extended discussion, consider the latest unclassified guidance on the subject from the U.S. Department of Energy:[9]

> The degree to which these obstacles [i.e., higher probability of predetonation in some designs, as well as increased heating and radiation compared with weapon-grade plutonium] can be overcome depends on the sophistication of the state or group attempting to produce a nuclear weapon. At the lowest level of sophistication, a potential proliferating state or sub national group using designs and technologies no more sophisticated than those used in first-generation nuclear weapons could build a nuclear weapon from reactor-grade plutonium that would have an assured, reliable yield of one or a few kilotons (and a probable yield significantly higher than that). At the other end of the spectrum, advanced nuclear weapon states such as the United States and Russia, using modern designs, could produce weapons from reactor-grade plutonium having reliable explosive yields, weight, and other characteristics generally comparable to those of weapons made from weapons-grade plutonium. . . . Proliferating states using designs of intermediate sophistication could produce weapons with assured yields substantially higher than the kiloton-range possible with a simple, first-generation nuclear device.

Even if reactor-grade plutonium can be used in weapons, it might still be possible to make access to it more difficult by technical or institutional means. Examples of the former are modifications of the standard Purex reprocessing flowsheet so that plutonium would not be separated from uranium or only partially decontaminated from fission products. Such schemes were assessed during the Carter administration and found not to offer significant nonproliferation advantages.[10]

The institutional schemes are more ambitious; their essence is to restrict weapon-usable materials and the technologies that produce them to international or multinational energy centers. The centers would take back spent fuel from national reactors and process it, along with inputs of natural uranium (and thorium in some schemes) to produce fresh fuel for both the national reactors and reactors located within the center. Many variants on this theme were also studied during the Carter administration, such as uranium-plutonium and uranium-thorium fueled pressurized-water reactors (figure 7).

Figure 7. Light-Water Reactor Fuel Cycle for International Safeguards, National Reactors Fueled with Thorium and Denatured Uranium

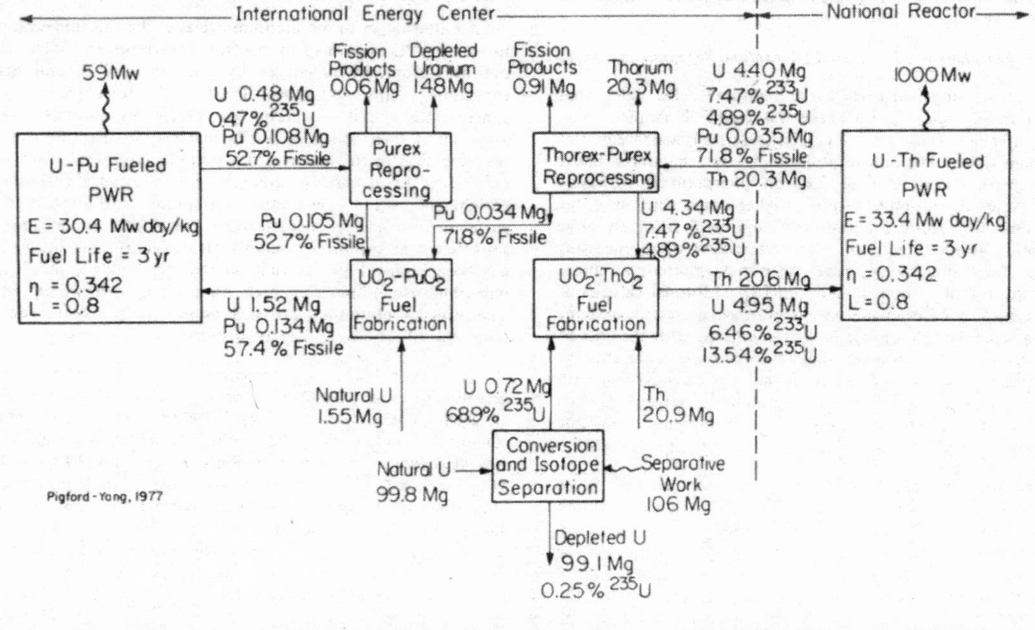

Source: "Report to the American Physical Society by the Study Group on Nuclear Fuel Cycles and Waste Management," *Reviews of Modern Physics* 50 (1, Part 2)(January 1978): S156.

Unsurprisingly, none of these schemes made it past the paper stage. The technical, economic, and institutional difficulties involved in setting up and operating such centers are considerable.[11] There is also a more fundamental problem: Who would sign up on the national reactor side of the proposed energy divide? Non-nuclear-weapon states would see such schemes as another means by which the weapon states seek to maintain their weapons monopoly. Are countries today more willing than they were in the past, when the perceived benefits of nuclear power were much greater, to embrace such discriminatory arrangements? This question is discussed further below.

In retrospect, the fears of the Carterites that the spread of nuclear power would lead to a "nuclear-armed crowd" did not materialize. Because of a number of factors, such as concerns about reactor safety, especially after the Chernobyl accident, the lack of progress in disposing of spent fuel, and the availability of cheap natural gas as feed for efficient, modular gas turbines, nuclear power has hardly spread beyond the countries that had already implemented it when Carter took office. During the administration of President Ronald W. Reagan the proliferation issue moved to the back burner, and its focus shifted from the fuel cycle to the efforts of Pakistan to develop a nuclear weapon capability via unsafeguarded dedicated facilities. As a result of the Chernobyl accident, technical innovation in nuclear power for the past fifteen years has focused largely on designing safer reactors.

At the same time, attempts to increase the proliferation resistance of the fuel cycle never entirely disappeared. For example, proponents of the integral fast reactor claimed that it not only was safer and produced less long-lived waste requiring geologic disposal than a standard fast breeder reactor, but also that its integral design and pyroprocessing technology made it more proliferation-resistant.[12] The major selling point of the thorium-uranium light-water reactor fuel invented by Alvin Radkowsky is the claim of increased proliferation resistance compared with a light-water reactor fueled with low-enriched uranium.[13] How valid are these claims, and can newer designs do even better?

Will a Second Nuclear Era Be More Proliferation-Resistant?

The mood of proponents of nuclear power in the United States is decidedly more bullish these days. There is an electricity crisis in California, and nuclear power has powerful friends not only in Congress but also in the new administration. Within the nuclear research and development community there is much discussion of and some funding for new reactor designs that are advertised as simpler, safer, and more proliferation-resistant.

Are such designs technically feasible, economically viable, and politically acceptable?

Technical Feasibility

Some of the new designs involve evolutionary, if still undemonstrated, extensions of current technology; others have important novel features and/or require the integration of several advanced technologies. Examples of the former are the high temperature gas reactor, the Radkowsky thorium-uranium-fueled light-water reactor, and the international reactor innovative and secure (a light-water reactor variant of the secure transportable autonomous reactor concept). Examples of the latter are the encapsulated nuclear heat source variant of the secure transportable autonomous reactor concept and the accelerator transmutation of waste concept proposed by the Los Alamos National Laboratory.

A common proliferation-resistant feature of the reactor designs is high burnup, long-life cores—on the order of 100,000 megawatt-days per metric ton and ten years, respectively—without fuel shuffling or refueling. This feature reduces the need for frequent access to the core in off-load designs and increases the fraction of even plutonium isotopes in the discharged fuel. In the Radkowsky design, the high burnup also increases the amount of uranium-232 produced along with uranium-233 in the thorium-fueled blanket.

The production of such cores requires considerable research and development. Experience with naval reactors is probably relevant, but the classification restrictions hamper the technology transfer.[14] It is also debatable how much the degraded isotopics contribute to proliferation resistance. As the previously cited Department of Energy guidance on reactor-grade plutonium implies, some nuclear-weapon designs are predetonation-proof; that is, they work with any isotopic mixture of plutonium. Such designs are certainly advanced compared with the Trinity bomb, but were developed many years ago, so that many states now have the capability to implement them. Although the excess heat from plutonium-238 in reactor-grade plutonium and the radiation from the uranium-232 daughters in thorium cycles must also be accounted for in weapon design and production, neither is considered to be a significant obstacle.[15]

The other side of the proliferation coin with regard to high burnup is the increased enrichment required at the front end, on the order of 10 percent for a burnup of 100,000 megawatt-days per metric ton. As is well-known, the use of such material instead of natural uranium as feed for a uranium enrichment plant significantly reduces the separative work required to produce weapon-grade uranium, and hence the size of the plant. Less well-known is that uranium-233 is an excellent weapon material and is easier to enrich to weapon-grade than uranium-235, particularly using centrifuges.

Economics

In the past, governments heavily subsidized the development of civilian nuclear power on the basis that it was important for both national prestige and security. The fact that it could also serve as a convenient conduit and cover for a nuclear weapon program was also widely appreciated. Given this, it was tempting to fudge the numbers with regard to the real cost of nuclear electricity. Now, in an era of increasing deregulation, nuclear power must meet the test of the marketplace. An important consideration is whether increasing the proliferation resistance of the fuel cycle will be a significant burden in this regard. Conversely, will new technologies designed to make nuclear power more competitive make it *less* proliferation-resistant?

Several examples come to mind, beginning with the rise and fall of laser isotope separation technology in the United States. In the 1960s it occurred to scientists in several countries, including the United States, France, and Israel, that it might be possible to use the then newly-developed tunable dye laser technology selectively to excite and then separate (enrich) isotopes of interest, e.g., the uranium-235 in natural uranium. Two means to this end were suggested, one using molecules and the other using atoms as the working material. These processes became known as molecular laser isotope separation and atomic vapor laser isotope separation, respectively. For uranium enrichment, the former used uranium hexafluoride, and the latter used uranium vapor. In the United States, government-sponsored programs to develop the molecular laser isotope separation and atomic vapor laser isotope separation processes for uranium enrichment started in the early 1970s at Los Alamos and Livermore National Laboratories, respectively. A private consortium, Jersey Nuclear Atomic Isotopes, also pursued atomic vapor laser isotope separation for uranium enrichment independently.

By the mid-1970s, in the expectation of a major expansion of nuclear power in the wake of the 1973 oil crisis, the Department of Energy was supporting research and development of four uranium enrichment processes—molecular laser isotope separation, atomic vapor laser isotope separation, the plasma separation process, and the advanced gas centrifuge—as well as a major upgrading of the large U.S. gaseous diffusion capacity. However, the expansion did not materialize, and the Department of Energy began to winnow the advanced options. In a 1982 process selection it chose atomic vapor laser isotope separation over molecular laser isotope separation and the plasma separation process for further development, and in a 1986 process selection atomic vapor laser isotope separation emerged the winner over the plasma separation process.

One consideration in the process selections was proliferation resistance. The author, along with Manson Benedict and George Rathjens, two colleagues from the Massachusetts Institute of Technology, were asked to make proliferation assessments of the competing technologies. As might be expected, proponents advertised that their processes would produce low-enriched uranium, suitable for nuclear reactors, at very low cost and could only be configured to produce highly enriched uranium, suitable for nuclear weapons, with great difficulty. The conclusion reached by the author, Benedict, and Rathjens was that a process that made low-enriched uranium could be modified to make highly enriched uranium with comparable difficulty. However, they did not dig deeply enough into the issue of whether it could make low-enriched uranium as advertised. The wake-up call, at least for some, came during the 1986 process selection, which the nuclear pioneer Karl Cohen characterized "as a contest between a technology that didn't work (AGC [advanced gas centrifuge]), and one that didn't exist (AVLIS [atomic vapor laser isotope separation])." Amidst all the hoopla, many people, including the author, had not appreciated that atomic vapor laser isotope separation as an integrated system existed mostly on paper; only miniscule amounts of enriched uranium had actually been produced. However, the Department of Energy and, after privatization of the enrichment program, the United States Enrichment Corporation continued their support of atomic vapor laser isotope separation until they pulled the plug in 1999.

The lesson here is that technology not only has to work on paper or in the laboratory. It must also be economic on a commercial scale. Similar pitfalls may lie ahead. For example, advocates of nuclear power using once-through fuel cycles point to seawater as a source of sufficient uranium to support thousands of gigawatts for hundreds of years. They make this assertion on the basis of extraction costs on the order of $200 per kilogram of uranium, estimated on the basis of recent experiments that recovered gram quantities of uranium from the sea off the coast of Japan. The argument is that further development will reduce these costs substantially, perhaps to $100 per kilogram of uranium. At current costs of about $25 per kilogram, uranium accounts for approximately 5 percent of the busbar cost of nuclear electricity, so that an increase to $100 per kilogram would increase the busbar cost by 15 percent. While this increase may appear tolerable, the world's nuclear vendors are unanimously of the opinion that nuclear power costs must be *reduced* by about 30 percent to be competitive with combined cycle gas turbines. Of course, the projected costs of uranium on the order of $100 per kilogram are far from assured; in fact, the latest estimate is about $1,200 per kilogram.[16] This discussion should not be taken as criticism of continued research and development in this area. In fact, such efforts should be increased to provide more realistic estimates of the recovery costs. In sum, it is unclear whether today's safer, simpler, modular, and more proliferation-resistant "paper" reactors and fuel cycles will turn out to be economically competitive.

What does seem clear is that even while greater proliferation resistance is desirable, it is not sufficient: The public is still concerned about safety and waste disposal, and rightly so. Moreover, innovation is not restricted to the nuclear fuel cycle. Both non-nuclear electricity sources, as well as opportunities for reducing energy demand via greater efficiency in end-use, will pose stiff competition to future improvements in the nuclear fuel cycle.

Political Acceptability

Last, but certainly not least, is the issue of the political acceptability of changes designed to increase the proliferation resistance of the nuclear fuel cycle. The discussion so far has focused mostly on new reactor designs as a means of increasing the *intrinsic* barriers to proliferation. However, the view that nuclear power generated by, for example, low-enriched uranium-fueled once-through cycles is an acceptable proliferation risk depends on the viability of both the *intrinsic* and the *extrinsic* barriers to proliferation, including safeguards, physical security, and export controls. Going beyond what might be called the "once-through cycle standard" for proliferation resistance logically involves upgrading extrinsic as well as intrinsic barriers. This process is ongoing—witness the recently negotiated additional protocol to the Nuclear Non-Proliferation Treaty model safeguards agreement designed to make a state's nuclear activities more transparent.

Whether such changes are sufficient is another matter. Those who think that they are not sufficient but who still want to retain nuclear power as a global, long-term energy option commonly invoke international or multinational energy centers or parks containing all sensitive nuclear activities as the ultimate solution to the proliferation problem. This approach raises doubts. As noted,[17] such arrangements would require that countries "outside the fence" restrict their peaceful nuclear activities. Already the non-nuclear-weapon states that are parties to the Nuclear Non-Proliferation Treaty are increasingly unhappy with the progress toward nuclear disarmament called for in Article VI of the treaty. Such new restrictions would run against the tide, not to speak of Article IV of the treaty. Alternatively, it is possible to imagine a nondiscriminatory regime in which all countries would accept international control of their nuclear energy programs, as proposed in the Baruch Plan of 1946, which was based in turn on the findings of the Acheson-Lilienthal Report. However, the Baruch Plan also required the destruction of all nuclear weapons. Indeed, it is hard to imagine a viable nuclear weapons–free world without international control of peaceful nuclear activities. This world seems as visionary today as it did in 1946, but it is probably the only way, short of abandoning nuclear power, to break the power/proliferation linkage.

Technical Opportunities for Increasing Proliferation Resistance of Nuclear Power Systems (TOPS) Task Force

James A. Hassberger

In 1999 the U.S. Department of Energy's Office of Nuclear Energy Science and Technology established the Technical Opportunities for Increasing the Proliferation Resistance of Nuclear Power Systems (TOPS) task force, a special unit of the Nuclear Energy Research Advisory Committee of the U.S. Department of Energy. The mandate of the task force was to identify and recommend both near- and long-term technical opportunities to enhance the proliferation resistance of global civil nuclear energy systems. Its work supported the program planning of the U.S. Department of Energy in this area, including by identifying long-term technology options. One premise underlying the work of the task force was that in the near term the United States and other developed countries were likely to continue to rely on nuclear power and that its use would expand globally.[1]

The TOPS process began with two workshops held by or in conjunction with the Center for Global Security Research at Lawrence Livermore National Laboratory.[2] One conclusion of the workshops was that, although the issues underlying technology for proliferation resistance had been under discussion for a long time, certainly since before both the Nonproliferation Alternatives Systems Assessment Program[3] and the International Nuclear Fuel Cycle Evaluation[4] studies, no concrete conclusions had ever been drawn regarding the specific role of technology for enhancing proliferation resistance. Over the last twenty or thirty years since these reviews, however, there have been significant advances in fuel and material technologies and in computers and instrumentation that might be brought to bear on the problem of proliferation in the near and longer term. The focus of the original workshop at Livermore was just that: to understand how technology might be applied to some of the proliferation-resistance issues and challenges that exist today. It began by looking in considerable detail at what some of these issues might be. Neither that review, nor the subsequent TOPS review, was meant to be a detailed technical evaluation.

The task force reached consensus on three broad points. First, there is both the potential and the need to pursue promising areas of research and development that are likely to enhance proliferation resistance. The TOPS task force did not single out any particular system. Rather, it concluded that there were a number of promising technical opportunities that merited investigation for their possible applicability and effectiveness.

The second point on which the task force reached consensus was that proliferation resistance and related technologies were only some of the important aspects of complete nuclear power systems that need further research and development. Others are the economics, safety, and waste disposal of the systems. Thus, proliferation resistance should be looked at in a broader context and weighed against other requirements. A fully proliferation-resistant nuclear power system achieved at the cost of safety, economics, or serious waste management issues would not be acceptable.

The third point of consensus was that other important opportunities exist for reducing the risk of proliferation worldwide. The task force singled out three key ones:

1. reduced risk of theft of weapon-usable materials, particularly in places like Russia and elsewhere in the former Soviet Union;
2. enhanced international safeguards systems (it is very interesting to note that the bulk of the money spent on international safeguards goes to safeguarding plants in weapons countries and very well-developed states. The efficacy of the funds would be far greater if the equation were reversed); and
3. continuing strengthening of controls over exports of technologies that can be used to produce nuclear weapons.

The TOPS task force also identified three important research and development objectives. The first (this order is not intended to reflect the weight or priority of the objectives) is a comprehensive evaluation of the nonproliferation implications of different options. Currently there is no systematic approach to evaluating either the risk of proliferation or the efficacy of options for reducing it. The task force concluded that it is very important to start pursuing mechanisms for conducting this evaluation.

The second objective is to explore and develop various approaches to improving proliferation resistance, which were generally stated as:

1. increasing the effectiveness and efficiency of institutional measures themselves, for example, by strengthening the safeguards;
2. making potentially usable materials highly inaccessible;
3. reducing the attractiveness of nuclear materials for use in weapons; and
4. limiting the spread of weapon-usable knowledge and skills.

All these approaches can contribute to proliferation resistance.

The final research and development objective is to evaluate the range of the technical options and fuel cycles that could engender broad international participation and maintain the desired level of performance with respect to safety, economics, and the environment.

As noted, the TOPS task force believed very strongly that the research and development objectives must be pursued in concert with meeting the other challenges and requirements facing nuclear power.

The TOPS task force found that the three research and development objectives translate almost directly into three broad research and development program areas. TOPS also considered three timeframes for achieving results: near, intermediate, and long-term. The first recommended program area involves the development of improved methodologies for assessing proliferation resistance. It is impossible to make reasonable decisions on which technologies, options, or fuel cycles are more or less proliferation-resistant without some basis for comparison. As an aid in its discussions the task force developed what has been called a barriers approach to proliferation resistance, which derives broadly from work of the National Academy of Sciences on this issue. The task force identified three categories of barriers to proliferation based on: the qualities of the nuclear materials; the technical impediments to their use; and institutional arrangements. These barriers are either intrinsic (inherent to the nuclear fuel cycle) or extrinsic (institutional). Any assessment of proliferation resistance must address both types of barriers, and the development of technologies should similarly aim to strengthen both.

It is widely agreed that the fissile materials used in commercial nuclear power systems are the key link to weapon applications. Civil nuclear power technologies, however, can also be used as a source of nuclear skills, knowledge, and expertise (that might be used in a nuclear weapon program); can facilitate the processing and use of the materials in a weapon program; and can help hide covert programs. Nevertheless, in its report and elsewhere the TOPS task force specifically noted that, historically, civilian nuclear power has not been the path of choice to fissile materials.

There are several barriers intrinsic to the nuclear materials used in commercial nuclear fuel cycle activities. The isotopic and chemical qualities of the materials do not make them ideal for use in nuclear weapons and are barriers to proliferation. The mass and bulk of nuclear materials are a very strong and effective natural barrier. Uranium ore is massive and bulky, as is spent fuel, which because of another intrinsic barrier (radiation) requires extensive and massive shielding. Radiation also complicates handling and/or processing, adds significant personnel hazards, generates heat, complicates transport, and so on. Nuclear material is, by nature, quite detectable, with different nuclear materials having different natural qualities of detectability. In short, all these factors increase the difficulty of using materials found in the nuclear fuel cycle relative to other paths to obtaining materials for a nuclear weapon program.

The technologies of nuclear power systems also offer intrinsic technical barriers. The facility may not be of a type that can be easily used to produce weapon materials, or whose fissile materials, new or spent, can be easily used for weapon development without a high level of knowledge and expertise. Access to the facility, the technology, and the materials may be inherently difficult. There is a question of whether the fissile materials in a facility are of sufficient mass to make an attractive target. The facility or process may have an intrinsic capability to detect abuse, theft, or misuse. There is a time factor: It takes time to modify a facility or gain access to materials under typical operating conditions. In short, there are also proliferation-resistant barriers in the nuclear power technologies.

Extrinsic barriers are developed and added to the nuclear fuel cycle to help control the remaining risks in the nuclear fuel cycle and compensate for possible limitations in the intrinsic barriers. Since it is not technically possible to eliminate all proliferation risks from the nuclear fuel cycle, extrinsic barriers such as treaties are necessary. For example, the United States demonstrated in World War II that it is possible to make a bomb out of "dirt" (that is, uranium ore), as the Manhattan Project did. Of course, the resources and technology required to do so are extensive. The question, therefore, is not what is zero risk, because there is none. Rather, it is how does one ensure that the civilian nuclear fuel cycle remains the least likely path for a potential proliferant's access to materials and technology required for weapon development.

The TOPS review of extrinsic barriers was limited to possible technological innovations that could strengthen the barriers, consistent with the study objective—to identify technology opportunities for improving the barriers. As such, the review excluded treaties, international standards, approaches to international control, and the like. While there are indirect ways in which technologies can help, such as by improving the ability to verify treaties, that connection was outside the charge given to TOPS.

The effectiveness of the extrinsic barriers depends heavily on how they are implemented. The task force looked at technological applications to safeguards, such as monitoring, detection, and deterrence of misuse of facilities and/or the diversion or theft of materials; and control of access and security, such as administrative controls, physical protection, and effective backup

As noted, the task force identified three important timeframes based on the point at which the improvement can be actualized, as opposed to when the research starts. The technical opportunities for reducing proliferation risk that can be realized in the near term, within, say, the next five years, are those involving the development and improvement of assessment methodologies. Although this assessment work would not directly improve proliferation resistance, it is needed to plan the development of actual technologies over the longer term. Thus this first stage is a very important aspect of implementing technologies. Another near-term program area is pursuit of technologies to enhance extrinsic barriers. Strengthening them provides a near-term benefit that can make a real difference with today's nuclear power systems without a lot of intrusive work involving the nuclear fuel cycle itself. Particular areas relating to enhanced extrinsic barriers that merit attention in the near term relate to institutional measures, improved safeguards, a better ability to monitor what happens, and improved information technology, all ideas that have been around for quite a long time. An example of one measure is improved information technologies, such as the integration of sensors and data systems, expert systems for data analysis, and real-time surveillance. Two others are the development of enhanced material tagging and monitoring, and improved wide-area monitoring, such as through remote surveillance. The International Atomic Energy Agency has programs in technologies such as remote video observation, which is unattended video that is beamed directly back to the agency. The next step, however, would be to connect that technology to the kind of imaging technology available today on some personal computers, so that the computer looks for differences in the images over time, with an automated early warning device that notifies staff when it detects differences. Such mechanisms could save the International Atomic Energy Agency real money and allow it to focus its efforts more efficiently.

The final near-term program area is assessment of the technical effectiveness of some of the intrinsic barriers. For example, minor modifications to the fuel cycle could be implemented now to increase the burnup of nuclear fuels. The assessment would look at the extent to which such a change could really improve proliferation resistance.

The intermediate-term technologies are those that might come to fruition within the next fifteen years. Here the focus would be on developing technologies to enhance intrinsic barriers and to improve existing systems and those near implementation. In the case of light-water reactors, attention would be paid to extending the burnup to reduce the quantity and quality of the plutonium, and using thorium cycles to further reduce its quantity and quality. With high-temperature gas-cooled reactors, the objective would be low fissile loading, very high burnup, and low production of plutonium. For fast reactors, the objective would be recycle-in-place systems that eliminate reprocessing and ultra-long-life fuels. Research and development with small reactor systems would emphasize fueled-for-life cores that eliminate all on-site fuel handling, and with advanced recycle systems it would be recycle systems that eliminate separable weapon-usable materials for both closed and transmuted fuel cycles.

The longer term options, which involve a timeframe beyond fifteen years, would include both the development of radically new fuel cycles and technologies that address proliferation resistance and other issues from a more fundamental approach, and an assessment of their implications. Even though classified as longer term options, the research and development would not necessarily be postponed for ten or fifteen years. Some options need enough early research and development to permit an initial understanding of their implications and to provide the information needed to determine how effective they might or might not be, since clearly not all will be effective.

The TOPS task force did not come up with specific technology recommendations because there are currently too many uncertainties associated with assessing how effective many of these technologies are for enhancing proliferation resistance. An example is the high-temperature gas-cooled reactor. It has the great advantages of very high burnup and superior thermal efficiency, so that it does not make as much plutonium per unit of electricity as other types of plants. Its spent fuel has very dilute plutonium: each of the billiard ball–sized fuel elements has only a few milligrams of plutonium, and almost a full core would have to be diverted to derive a significant quantity of plutonium. On the other hand, because there are thousands of the little balls, material accountability may be a bit more difficult.

There are a lot of issues that have to be balanced against one another, and the weight assigned to each issue, based on their attributes and advantages and disadvantages, depends a lot on who is doing the talking. As such, a better understanding of the implications of these technologies, and what their limits and prospects are, is needed, as is a framework for making those evaluations.

The task force did arrive at four principal conclusions. First, there are promising technology opportunities for enhancing proliferation resistance but no good mechanism for evaluating them systematically. Second, early evaluation of the wide range of proposed fuel cycles and options is critical to developing a research and development plan that is affordable and realizable. Third, proliferation-resistance enhancements can only be accomplished if economic, safety, and environmental objectives are also achieved. Fourth, credible research and development must be done to establish the potential value of the various options.

Disclaimer

necessarily constitute or imply its endorsement, recommendation, or favoring by the United States government or the University of California. The views and opinions of authors expressed herein do not necessarily state or reflect those of the United States government or the University of California, and shall not be used for advertising or product endorsement purposes.

10

The Limits of Technical Fixes

Edwin S. Lyman

The global commercial nuclear power industry has generated well over a thousand tonnes[1] of weapon-usable plutonium. Although most of this plutonium remains bound in highly radioactive spent nuclear fuel, about 200 tonnes that have been chemically extracted through reprocessing are stored in separated form worldwide. The proliferation issues associated with the separation, processing, and transport of plutonium are widely known (although their significance remains controversial). However, the risks associated with the large-scale accumulation of unseparated plutonium have received comparatively little attention, except in politically charged circumstances, such as Russia's supply of reactors to Iran. When considering the question of whether nuclear power should have a role in future energy generation, both aspects must be considered.

There is a growing acknowledgment of the proliferation risks posed by the nuclear power industry, whether or not plutonium is separated. The notion that future nuclear energy systems must be more proliferation-resistant than the current generation has become practically gospel in some quarters. The U.S. Department of Energy and Russia's Ministry of Atomic Energy have both embraced the concept, with the latter's views apparently receiving endorsement at the highest levels of the Russian government. Less enthusiastic about this program are the nations that currently engage in reprocessing, including France, the United Kingdom, and Japan, since any suggestion that nuclear power systems must be made more proliferation-resistant in the future carries the implication that the risks they pose are too high today. Nonetheless, even Director General Mohamed ElBaradei of the International Atomic Energy Agency, which is directly responsible for ensuring the effectiveness of the safeguards system in states that are members of the Nuclear Non-Proliferation Treaty, concedes that "the future of nuclear energy may depend heavily on success in developing new, innovative reactors and fuel cycle designs that exhibit enhanced safety features, proliferation resistance and economic competitiveness."[2]

The renewed attention to the proliferation risks associated with nuclear power is a welcome development. However, the motivation of some advocates of proliferation resistance needs to be examined more closely. In the United States nuclear nonproliferation policy is still nominally guided by the 1993 statement from the administration of President William J. Clinton, which says that "the U.S. does not encourage the civil use of plutonium and, accordingly, does not itself engage in plutonium reprocessing for either nuclear power or nuclear explosive purposes."[3] This policy has been a perennial source of frustration for plutonium enthusiasts within the Department of Energy, who saw numerous projects—such as Argonne National Laboratory's integral fast reactor—canceled on this basis.

Clearly, one way of bypassing the letter (although, arguably, not the spirit) of the policy would be to develop plutonium separation techniques with so-called "proliferation-resistant" features not considered to be "reprocessing" under a strict definition of the term—that is, without a complete separation of weapon-usable materials from highly radioactive fission products. This reasoning was apparent when former Minister Yevgeny Adamov of the Ministry of Atomic Energy proposed a moratorium on civil reprocessing at the 1999 International Atomic Energy Agency General Conference—a moratorium not intended to apply to the development of a new fleet of plutonium-fueled fast breeder reactors based on a "proliferation-resistant nuclear fuel cycle of natural safety."

The Technological Opportunities to Increase the Proliferation Resistance of Global Civilian Power Systems (TOPS) report, issued by a task force of the Department of Energy's Nuclear Energy Research Advisory Committee, made an indiscriminate recommendation that the Department of Energy initiate an expensive new program to pursue analysis, research and development, and, ultimately, demonstration of a wide variety of advanced nuclear power and fuel cycle concepts, including once-through thorium fuels for light-water reactors, "dry" (non-aqueous) reprocessing of light-water reactor fuels, liquid-metal fast reactors, and spent fuel transmutation technologies.[4] (The TOPS task force apparently never met a reactor system it did not like.) Accordingly, the Department of Energy is establishing the framework for a renaissance of government-funded nuclear energy research through its Nuclear Energy Research Initiative, Advanced Accelerator Applications, and Generation IV programs. All these items are targeted for big budget increases in the energy bills now before Congress.

The prospect of a new era of government-subsidized reactor and fuel cycle development, unfettered by the irritating constraints of nonproliferation policy, has led to a virtual feeding frenzy among national laboratories, moribund academic nuclear engineering departments, reactor vendors, and other government contractors. The Department of Energy's new initiatives have already spawned a number of truly bizarre reactor concepts. Grandiose new "architectures" for nationwide or worldwide nuclear fuel cycles have been proposed. What is missing so far is any interest on the part of electric utilities in these initiatives. A reality check is clearly called for.

To the extent that this newfound interest in proliferation resistance is merely a public relations ploy to gain acceptance of nuclear fuel cycles based on plutonium recycling—a nonproliferation seal of approval, as it were—it will be extremely counterproductive. To have a truly proliferation-resistant closed fuel cycle, the risk should be no greater at any point in the process than the risk posed by the once-through cycle—in other words, the process materials should meet the "spent fuel standard" at all times. Achieving this standard will not be a simple task and will likely raise the costs and health risks of the technology to unacceptable levels. However, anything short of this standard will at best provide only a marginal reduction in risk in exchange for an undoubtedly considerable cost. At worst it will provide a false confidence that will greatly increase the future danger of diversion or theft of weapon-usable materials.

The Irreducible Proliferation Risk of Nuclear Power

To understand the objectives of the technical fixes that have been proposed to address the proliferation risk of nuclear power, it is necessary to understand in more detail the nature of the problem they are trying to solve. As the National Academy of Sciences has pointed out, the concept of proliferation risk is highly dependent on the type of threat under consideration (national, subnational, or subnational with the support of a foreign state). For simplicity, here the focus is on the subnational threat.

A typical low-enriched uranium spent fuel assembly from a pressurized-water reactor contains about 4–5 kilograms of reactor-grade plutonium—roughly one bomb's worth—intimately mixed with about 450 kilograms of low-enriched uranium and other radioactive materials. The assembly weighs about 650 kilograms and is about 3 meters long. While the size, weight, and plutonium dilution provide some measure of protection against casual theft of a spent fuel assembly and ready conversion into a component suitable for a weapon, the most important barrier is the "self-protecting" penetrating radiation field. Ten years after discharge from the reactor, this field is typically on the order of 20,000 rem per hour at the surface and 1,500 rem per hour at a distance of 1 meter, compared with an acute lethal dose of about 600 rem. This barrier provides a considerable deterrent both to the theft of the spent fuel element (requiring access to a shielded shipping cask and the means to load and transport it) and to recovery of the plutonium (requiring access to a remote-controlled, heavily shielded industrial plant).

International (International Atomic Energy Agency) and domestic (Nuclear Regulatory Commission and Department of Energy) standards for physical protection and safeguards of direct-use, special nuclear materials (for example, plutonium and highly enriched uranium) are far less stringent for materials that are irradiated. For instance, fresh mixed-oxide fuel assemblies, each of which typically contains 20–30 kilograms of plutonium, must be inspected once a month, whereas irradiated fuel assemblies (either low-enriched uranium or mixed-oxide) must be inspected every three to four months. Moreover, the material accountancy protocols are more rigorous for unirradiated material. With regard to physical protection, irradiation of an item reduces the "category" by one grade and accordingly allows a relaxation of the requirements recommended by the International Atomic Energy Agency Convention on Physical Protection.

An inexorable fact of physics is that the radiation barrier of spent fuel decreases with time. Ten years after its discharge from the reactor, the penetrating gamma field of a spent fuel assembly is dominated by the isotope cesium-137, which has a half-life of thirty years. Eventually, the radiation barrier will decline to the extent that it can no longer provide a reasonable level of self-protection.

How much self-protection is enough? It is impossible to provide an objective numerical value because any reasonable definition is heavily dependent on the scenario under consideration and a host of additional assumptions. Nevertheless, the International Atomic Energy Agency has defined the lower limit of the dose rate for "irradiated" material for purposes of physical protection as 100 rem per hour at a distance of 1 meter. The Nuclear Regulatory Commission and Department of Energy also employ that standard.

A spent fuel assembly with a 1,500 rem per hour dose rate at 1 meter ten years after discharge will fail to meet the 100 rem per hour standard after an additional one hundred and fifteen years. If the assembly is in above-ground, interim storage at that time, consistent application of the rules would require an upgrade of physical protection at the facility. For example, special nuclear material considered "irradiated" under the above standard is exempt from a number of Nuclear Regulatory Commission physical protection regulations that would have to be applied once the radiation barrier fell below the threshold. (For this reason the Department of Energy is seeking an exemption from the 100 rem per hour limit for its surface spent fuel transfer facility at Yucca Mountain, which will require a Nuclear Regulatory Commission license.)

One way to mitigate the risk posed by spent fuel that is no longer self-protecting is to emplace and seal it in a geologic repository. The additional time and effort required to retrieve buried fuel can be regarded as a geologic barrier replacing the missing radiation barrier. At that point safeguards and physical protection requirements could be relaxed, although a modest level of security will most likely have to be applied indefinitely.

However, the fact remains that a geologic spent fuel repository will contain an enormous amount of weapon-usable plutonium, typically enough for hundreds of thousands of bombs, as well as substantial quantities of other long-lived weapon-usable isotopes, such as neptunium-237. A number of observers have dubbed these facilities "plutonium mines" and argue that because the material in repositories can be accessed using conventional mining techniques, the proliferation risk of burying plutonium is unacceptably high. While those espousing this view primarily employ it in a self-serving way to justify the development of spent fuel transmutation schemes (see below), the issue cannot be discounted entirely.

This issue was analyzed by Lyman and Harold Feiveson, who argued that the risk of theft of plutonium from a sealed geologic repository is meaningful only when compared to the risk that weapon-usable material could be acquired from other sources, such as operating fuel cycle facilities or clandestine production plants.[5] They concluded that the attractiveness of plutonium mining would be relatively low and that repository mining would be easier to detect than would other unauthorized activities such as clandestine reprocessing or uranium enrichment.

Nevertheless, these plutonium-loaded facilities will present risks that cannot be entirely eliminated. Moreover, if public opposition makes siting of underground facilities impossible, as could well be the case, spent fuel would remain in above-ground storage indefinitely, where the institutional measures essential to adequate protection cannot be guaranteed. Hence it must be acknowledged that the operation of conventional nuclear power plants is creating a massive plutonium inventory that will pose some risk far into the future. Before a large-scale revival of nuclear power can be considered, it must be determined decisively whether it is realistically possible to control this risk at an acceptable cost and without greatly increasing near-term environmental, public health, and proliferation risks. If the answer is negative, then policymakers should be prepared to concede that nuclear power indeed has no future.

What Does Proliferation Resistance Mean, and Can It Work?

The concepts of proliferation resistance fall into three categories.[6] First are those that involve increasing the proliferation resistance of the once-through fuel cycle by reducing the quantity or quality of weapon-usable material in spent fuel, thereby lowering the security burden associated with the storage and disposal of these fuels. Related approaches include thorium-uranium fuels, inert matrix (uranium-free) fuels, and ultrahigh burnup gas-cooled reactor fuels.

The second category are those that involve developing processes for plutonium recycle that pose smaller proliferation risks than the aqueous reprocessing and mixed-oxide (plutonium-uranium) fuel fabrication processes now employed to varying degrees by France, the United Kingdom, Japan, and Russia. These processes would not require complete separation of plutonium from fission products or other actinides, so that the reprocessing product and the recycle fuel would always retain some measure of radiological self-protection.

The two categories are linked in the following way. For technical reasons, development of a once-through system capable of effectively reducing the quantity of weapon-usable material in the spent fuel to a level below concern is an unlikely prospect. In theory, only systems that involve repeated reprocessing and recycling would be able to achieve this. However, the uncertain benefit of long-term risk reduction would not be worth the near-term proliferation risks associated with large-scale reprocessing unless those risks could be decreased substantially.

Perhaps an even more important factor is that the intrinsic enhancements in the proliferation resistance of closed fuel cycles have to be so robust that existing safeguards and security requirements can be reduced with confidence. Safeguards and physical protection are expensive. As members of the International Atomic Energy Agency's

Safeguards Inspectorate are fond of mentioning, the agency's safeguards budget has not increased in real terms in more than a decade. It is unclear where the funds would come from to pay for the additional safeguards activities necessary to support a large-scale expansion of nuclear power, especially for a fuel cycle involving bulk handling of weapon-usable materials. Ultimately, the cost of safeguards should be internalized in the cost of nuclear energy generation. However, no matter who ends up paying for safeguards—the taxpayer or the ratepayer—proliferation-resistant features will not be of any practical benefit unless inspection resources and physical protection costs can be drastically cut. It is unlikely that this goal will be achieved with any of the technologies now under discussion.

The application of safeguards for low-enriched uranium-based, once-through cycles is considerably simpler than for closed fuel cycles based on reprocessing and recycle. The difficulty of implementing effective safeguards increases dramatically as the focus shifts from simple item-counting, tags, and seals to material accountancy at bulk-handling facilities, where statistical errors and biases in measurement techniques can create large uncertainties that can conceal diversions of material. These uncertainties are already unacceptably large at bulk-handling facilities today, such as reprocessing and mixed-oxide fuel fabrication plants, even though there are opportunities at some process stages for fairly accurate and precise assays of pure plutonium materials. The dirtier the material, however, the less precise and accurate the assays will become. Some argue that this reduction in the capabilities of material accountancy would be tolerable because the diversion threat would be reduced as a result of the reduced accessibility of the materials, so that containment and surveillance would provide adequate assurance against diversion. However, any system will have diversion pathways that can defeat containment and surveillance. There is no substitute for material accountancy.

Both categories of systems described above share a similar vulnerability—their proliferation-resistant features depend on the systems being operated as designed. Continuous direct inspections will be needed to ensure that unauthorized modifications are not being made to the system, with special attention required during refueling or maintenance outages.

This fact has led to proposals for a third category of systems—modular, long-lived "nuclear batteries" that contain the nuclear fuel and primary coolant system and are designed never to be opened during their lifetimes. Proponents of these systems argue that the batteries could be shipped to balance-of-plant sites all over the world, including to nations of proliferation concern, since tampering with the system would be difficult and readily detectable.

From a nonproliferation point of view, the least objectionable technologies are in the first category. Trying to improve fuel utilization and to reduce the attractiveness of spent fuel for weapons without the need for reprocessing is a reasonable goal. However, it is necessary to take a hard look at whether the modest benefits of these approaches are worth the considerable costs of development and deployment.

Reducing the amount of plutonium or other weapon-usable materials in spent fuel sounds like a good idea, given the long-term proliferation risks of spent fuel accumulation. However, the concepts proposed in this area are predicted to achieve only about a fivefold reduction in plutonium concentration,[7] whereas a system would have to achieve a much greater reduction—probably on the order of a factor of one hundred—before it would have a meaningful impact. One example shows why. A useful unit for considering the threat of theft is the maximum amount of spent fuel contained in a typical shipment (that is, to a reprocessing plant, an interim storage facility, or a repository). Current generations of spent fuel shipping casks can carry as much as 10 tonnes of spent fuel, and a given shipment can contain several such casks. A shipment of four casks would contain approximately one hundred bombs' worth of plutonium. Therefore, a two order-of-magnitude reduction in the plutonium content of spent fuel would be needed before such a shipment would cease to be attractive to subnational terrorist groups seeking a single nuclear device. In the long term the attractiveness of the plutonium inventory in a Yucca Mountain–sized repository (on the order of two hundred thousand weapons) or a large interim storage facility (on the order of ten thousand weapons) would barely be affected by anything short of a hundredfold reduction—and even that would be inadequate to alleviate the concern.

An even less useful characteristic of these proposed technologies is the reduction in plutonium quality that proponents claim they could achieve—that is, an increase in the percentage of heat- and radiation-emitting plutonium isotopes such as plutonium-238 and plutonium-240. It was to be hoped that this tired old argument would have been retired by now. It is well-established that plutonium of any isotopic composition is weapon-usable. Perhaps the most striking public statement on this point was made during a workshop at Lawrence Livermore National Laboratory in June 1999. Bruce Goodwin, a prominent weapon scientist at the laboratory, introduced the concept of "explosively fissile materials" to include "any fissionable material that can be assembled such that an explosive disassembly is possible."[8] He included as explosively fissile materials not only plutonium metal of any isotopic composition, but also other actinides, including americium-241 and neptunium-237, as well as oxides and other compounds containing these isotopes. Concerning the technical challenges associated with using explosively fissile materials in nuclear weapons, Goodwin stated that "[the] technical challenges can usually be overcome with engineering solutions" and that "experience has shown this to be true in the case of Pu [plutonium] and U [uranium]."[9] Moreover, "as nuclear weapon design and engineering expertise combined with sufficient technical capability become more common in the world, it becomes possible to make nuclear weapons out of an increasing number of technically challenging explosively fissile materials, many of which are components of spent reactor fuel."[10]

This statement leaves little room for doubt that the United States, and presumably other advanced nuclear states, have overcome these technical challenges and are capable of utilizing in nuclear weapons not only plutonium of any isotopic composition, but also other isotopes that present even greater obstacles. Confidence in a nuclear power system that slightly degrades the isotopics of plutonium in spent fuel would be contingent on the belief that this weapon expertise would not eventually diffuse to other nations and subnational groups. Given Dr. Goodwin's statements, this does not appear to be a safe bet.

The same arguments apply to pyroprocessing, a reprocessing technology that proponents claim is proliferation-resistant because it does not involve the separation of plutonium from other actinides. According to Dr. Goodwin's statement, many of those other actinides are weapon-usable as well.

Proponents have also oversold the effectiveness of the radiation barrier provided by residual fission products in the pyroprocessing system. Lyman, for instance, showed that the radiation barrier of the recycled fuel would be well below the 100 rem per hour criterion at the time of fabrication and would continue to decrease rapidly with time.[11]

The Problems with Proliferation-Resistant Technologies

The TOPS report asserts that the development of proliferation-resistant technologies "could be pursued under terms that are fully compatible with the need to assure that nuclear power continues to adhere to rigorous safety and environmental standards" and, moreover, that "many of the options would appear to be compatible with the objective of assuring that nuclear power is competitive with alternative energy sources."[12] There may be examples where this is the case, but no one has yet identified them. In fact, many features designed to enhance proliferation resistance conflict directly with the goals of increasing occupational safety and protecting the environment and public health. In addition, there is absolutely no evidence that any of the options proposed so far would be economically attractive.

There are numerous safety and environmental issues associated with proposed proliferation-resistant systems. A short list includes:

1. *Safety of ultrahigh burnup fuels and long-lived cores.* A trend toward ultrahigh burnup fuels can increase the radionuclide source term in the event of a severe accident. Moreover, the materials technology for such fuels is not at hand. Surprises continue to occur with fuels with the burnup levels encountered today.

   The same concerns apply for long-lived core concepts—especially those that are designed not to be opened by the recipient country. It is not credible or possible, considering the state-of-the-art in materials science and nuclear technology, to develop a system that could be operated safely for a ten- or fifteen-year period without any need to access the core for inspection or for emergencies. At a recent meeting, one of the Nuclear Regulatory

Commission's regional administrators observed, in regard to these proposals, that in his experience the most expensive nuclear plant systems to repair are the ones that were not intended to be replaced. The nuclear industry continues to have unanticipated problems, even with systems and materials that they believe they thoroughly understand. For some of the proposed new systems, which would utilize fuel and coolant materials for which there is little accumulated experience, unpleasant surprises are a certainty.

There will always be a tension between the security and safety goals at nuclear plants. They experience that tension today. In some cases, increasing the security of a plant's vital systems against potential saboteurs requires increasing the delay time for access to certain areas—exactly the opposite of the requirement that operators have facilitated access in an emergency. Striking the right balance is a challenging task.

2. *"Dirty" reprocessing and recycle.* As discussed, there is one necessary (albeit insufficient) condition for ensuring that a reprocessing and recycle system does not pose a greater proliferation risk than the once-through cycle: Any weapon-usable material within the process should never be more accessible than the plutonium in the initial spent fuel. That is, the spent fuel standard should apply throughout the cycle. None of the proposed systems meets this standard. Moreover, the logic of self-protection—to modify systems to increase their hazards to workers and the public so that they deter access—clearly conflicts with basic principles of industrial hygiene and environmental protection. This point should give pause to proponents of these systems. It is not obvious that the world really needs nuclear energy systems that must be operated in as dangerous a manner as possible in order to be considered proliferation-resistant.

Economic Considerations of Proliferation Resistance

Economic considerations are essential. In spite of the claim of the TOPS task force that proliferation resistance can be economically attractive, current trends appear to contradict that view.

There is a clear evolution in electricity generation away from large-capacity baseload plants toward small-capacity, widely distributed modular plants in both the developed and developing worlds. Modular gas turbines are the prime example of this type of system. Because of this shift, some in the nuclear industry believe that nuclear power systems will be competitive sources of electricity and heat generation in the future only if they can be designed to imitate the favorable characteristics of modular gas turbines. They may be right. However, nuclear power simply may not be amenable to this kind of deployment for both safety and nonproliferation reasons. A reduction in the proliferation risk can be most easily achieved by centralizing nuclear facilities and materials in a small number of sites, an approach that would reduce the resources necessary to safeguard and protect them and that would minimize the number of transport links required. To support a broad deployment of small nuclear plants based on a gas-turbine distribution model would require a highly credible means of reducing the associated proliferation risks, without at the same time causing unmanageable strains on the safeguards system.

It is not evident that the modular systems proposed to date meet this test. For instance, one particularly risky concept—modular, plutonium-fueled, liquid-metal cooled fast reactors with long-lived cores—clearly is vulnerable to interception during transport. The designer claims that proliferation resistance would be achieved during shipping by embedding the plutonium fuel in solid lead coolant. However, given the low melting point of lead, this barrier hardly provides significant protection if the entire shipment is hijacked.

Potential Showstopper: The Threat of Radiological Sabotage

Perhaps the greatest obstacle to a greatly expanded deployment of nuclear power plants, especially to politically unstable regions, is the threat that the reactor will become a target of radiological sabotage. An armed assault on a nuclear plant's vital safety systems can result in a core melt, a failure of containment, and a massive, Chernobyl-like release of radioactive materials into the environment. Radiological sabotage is a prime example of "asymmetric" warfare: The injury and property damage that a quantity of high explosives small enough to fit into a backpack can cause can be magnified a thousandfold if applied strategically at a nuclear plant. Such an assault could conceivably fulfill the same goals for a terrorist group as the acquisition and use of a crude nuclear weapon.

Recent experience in the United States, which has one of the world's most rigorous sets of requirements for physical protection at power reactors, has graphically demonstrated the challenges inherent in defending a nuclear plant against sabotage by armed attackers. The Nuclear Regulatory Commission has a program called the Operational Safeguards Response Evaluation that uses force-on-force exercises to test the effectiveness of the security at nuclear power plants. About 50 percent of the plants tested under the program have failed, meaning that the mock attackers would have been able to disable enough plant systems to cause "significant core damage." This statistic has not improved over the course of the program. The high failure rate occurs even though the exercises are scheduled well in advance and nuclear plant security forces undergo significant preparation, conditions that would not be present in a real terrorist attack. Moreover, the capabilities of the attacking forces in the exercises are often artificially constrained to the extent that it is questionable whether they accurately represent the capabilities of real-world adversaries.

The lesson of the U.S. Operational Safeguards Response Evaluation program is that significant resources, manpower, and training are required to defend nuclear plants against radiological sabotage. U.S. nuclear plant operators consider the costs of maintaining the security programs necessary to pass the operational safeguards test to be burdensome, and they have been actively seeking reductions in the physical protection requirements. In any case, it is highly doubtful that providing adequate physical protection for a system of widely distributed nuclear plants, some of which may be close to urban areas, will be affordable. A large part of the economic argument for modular reactors is the ability to reduce plant staffing significantly, but this is likely to be impeded by the security force requirements. Even nuclear plants that are fully proliferation-resistant will remain vulnerable to sabotage, and even the safest designs will have weak links to exploit. A further threat is sabotage of shipments of nuclear materials, a threat that is currently underplayed but clearly needs to be assessed.

The risk of radiological sabotage is of particular concern in politically unstable regions in the developing world, which are among those targeted as the most likely customers for small, modular nuclear plants. For instance, Indonesia has shown great interest in the gas-cooled pebble-bed modular reactor, a proposed new type of reactor that would not have a robust containment and would be uniquely vulnerable to sabotage attacks. In 1999 it presented the results of a study that surveyed the entire archipelago—from Aceh to Irian Jaya—for candidate sites for reactor placement.[13] Considering the level of ethnic violence and corruption in that country, it is legitimate to question the possibility of finding security forces that can be trusted to guard a large number of plants in the densely populated islands.

Accelerator Transmutation of Waste: A Totalitarian Scenario

A key issue is whether a technical solution can be found to the problem of nuclear waste and plutonium accumulation that will free nuclear power from the nagging environmental and proliferation problems that plague it. With geologic repository development on a slow track all around the world, some researchers have stepped into the breach with proposals for ambitious spent fuel transmutation schemes—elaborate systems, either reactor- or accelerator-based, that promise to rid the nuclear fuel cycle of all long-lived radioactive wastes, including plutonium and other weapon-usable actinides.[14]

One of the more prominent proposals, arising from the Los Alamos National Laboratory, is known as accelerator transmutation of waste. In recent years the Department of Energy has provided significant funding for transmutation studies, with support from the powerful New Mexico Republican Senator Pete Domenici, a staunch supporter of the laboratory, as well as from congressional opponents of the Yucca Mountain repository. In 1999 the department issued a report to Congress entitled "A Roadmap for Developing ATW [Accelerator Transmutation of Waste] Technology."[15] The report describes how a massive nationwide system of spent fuel reprocessing plants, accelerator-driven spallation neutron sources, liquid-metal cooled transmutation target assemblies, and pyrochemical reprocessing plants—a prime example of what some have referred to as a new fuel cycle architecture[16]—could transmute the entire U.S. spent fuel inventory over a 118-year period for a cost of only $279 billion (1999 dollars).

It may well be true, in theory, that such an architecture is the only way to manage the problem of plutonium accumulation. However, the transmutation roadmap does not address who is going to pay to design, build, and run these facilities for more than a century—not to mention protect and safeguard all the nuclear material processing and transportation—and do so with the degree of rigorous coordination essential for the slightest chance of success. One of the basic facts of transmutation is that if the system fails before the job is complete, the actual reduction in plutonium and other wastes would be less than an order of magnitude, and the project would be nothing more than a big waste of money and time.

To the question of who would pay there is only one answer—the government. Given the chaos plaguing the electricity industry today, it is very difficult to imagine that it would ever be able to act as one coordinated unit unless it were nationalized. As the roadmap report itself concedes, "it is unlikely that the private sector would implement a waste transmutation scheme on its own without incentives. . . . The federal government would have to play the primary role in organization, management and funding of any such system."[17]

Such an eventuality is not impossible. If society decides that ridding the country of plutonium and other long-lived nuclear wastes is a public good worthy of massive taxpayer support, then a government takeover of electric utilities for this purpose may be justifiable. However, the public—not to mention the conservative members of Congress who are staunch supporters of the program—must be made aware what it would take to solve the problem and should be fully informed of the costs and risks involved. If neither the public nor the private sector decides that it is willing to subsidize such an effort, then the nuclear industry must be willing to accept that its continued operation is unsustainable.

PART III

The Role of Nuclear Power in the Acquisition of Nuclear Weapons

11

Overview of Nuclear Power and Nuclear Weapons

Zachary S. Davis

The nuclear nonproliferation regime is a combination of domestic laws, international institutions, technical arrangements, and bilateral agreements, all of them held together by skillful diplomacy and a little smoke and mirrors. The focus here is on the effectiveness of the technical and legal barriers established to maintain the separation between civil and military applications of nuclear technology.

This issue is approached in two ways. The first entails case studies of three countries, India, Pakistan, and Iran, by Dr. George Perkovich, at that time with the W. Alton Jones Foundation. India provides a useful retrospective look at how civil nuclear technology contributed to its nuclear weapons program. When looking back on the history of the nuclear nonproliferation regime, it is useful to ask several questions. How effective were the instruments that were put in place? How effective were the U.S. nonproliferation laws and policies with respect to India? What was the net effect of the conditions put on the supply of nuclear technology enshrined in the Atomic Energy Act, Nuclear Non-Proliferation Act, and other important legislation intended to constrain India's nuclear weapons program? Did U.S. laws and policies successfully delay and defer New Delhi's decisionmaking regarding its use of civil nuclear technology to build nuclear weapons? Without U.S. restrictions, where would India's nuclear weapons program be today?

Some suggest that U.S. nonproliferation policy merely antagonized India without stopping its nuclear bomb program. Critics point to the sanctions policy put in place after the May 1998 tests as proof of the ineffectiveness of U.S. nonproliferation efforts with India. Instead, they argue, that policy only impeded efforts to improve relations with a prospective ally and fellow democracy. From this perspective nonproliferation is more trouble than it is worth. Some would even have the United States accede to India's longstanding request for transfers of nuclear technology, even though such transfers would require major revisions of U.S. nonproliferation laws and policies. Others, however, give credit to U.S. nonproliferation policy for limiting the pace of proliferation in South Asia and elsewhere, and view special treatment for India and Pakistan as further undermining the regime.

Cracks are developing in the nonproliferation regime's practices and institutions, particularly with respect to the policy of the Nuclear Suppliers Group to require full-scope safeguards as a condition of supply. Russia is disregarding this requirement so that it can sell reactors to India. What are the long-term consequences of violations of the full-scope safeguards standard for international behavior? How would the breakdown of the standards of the Nuclear Suppliers Group Guidelines affect proliferation elsewhere?

The case study of Iran forces us to look forward. Will Iran follow India's example and use its civil nuclear technology infrastructure to develop nuclear weapons? Iran is different in that it is a party to the Nuclear Non-Proliferation Treaty, but the direction in which it moves may raise questions about the long-term health of the nonproliferation regime and the treaty itself. Do the cracks in the regime make it easier for Russia, and perhaps other countries, to provide sensitive nuclear technology to Iran? If Iran diverts civil nuclear technology or material to military uses, will the International Atomic Energy Agency detect it, and will the members of the Nuclear Non-Proliferation Treaty take enforcement action?

The essay of the Honorable Lawrence Scheinman, professor emeritus, renowned scholar, and former assistant director of the Arms Control and Disarmament Agency, on the role of nuclear power with respect to nuclear weapons focuses on the health of the nuclear nonproliferation regime itself. Is the nuclear nonproliferation regime in trouble? Is it likely to unravel? Are the legal, multilateral, and technical barriers put in place to maintain the separation between civil and military nuclear activities wearing thin? If the regime is in decline, how long will it take before it loses credibility—five, ten, or twenty-five years?

Scheinman is one of the world's leading scholars on the International Atomic Energy Agency, a unique institution whose international safeguards system is an essential component of the nonproliferation regime. The answers to many of the questions raised about nuclear nonproliferation depend on the effectiveness of the agency's safeguards. Will the International Atomic Energy Agency meet the challenges posed by the states outside the Nuclear Non-Proliferation Treaty—India, Israel, and Pakistan—and by the states within it, such as North Korea, Iran, and Iraq? How will the agency handle the challenges posed by growing stockpiles of fissile material, the central issue of this book? Is the agency up to the job?

A number of reforms have been instituted to revitalize the International Atomic Energy Agency and make the safeguards system stronger, more robust, more intrusive, more reliable, and more credible. Nevertheless, questions persist about the so-called 93 + 2, or strengthened, safeguards system (formally known as the Additional Protocol to State Safeguards Agreements). With the agency's meager budget already stretched to the breaking point, how can it be expected to expand the coverage of its safeguards to include more fissile material, more facilities, and a wider range of activities?

In addressing this and other nuclear nonproliferation issues, some experts point to the nuclear bargain between the nuclear weapon states and developing countries, a two-part bargain. First, the nonweapon states are promised access to civil nuclear technology as part of their pledge not to acquire nuclear weapons. Second, the nuclear weapon states promise to reduce their arsenals and move in the direction of disarmament. Expectations, however, are outpacing the results from the nuclear bargain, and the stresses are becoming increasingly evident at Nuclear Non-Proliferation Treaty review conferences. They are evident in the threats of some countries to quit the treaty. Are these threats serious? Is the treaty in trouble?

The challenges to the safeguards regime do not give cause for optimism. In light of the demands being placed on the International Atomic Energy Agency, the funding shortfalls, the strains outside and inside the Nuclear Non-Proliferation Treaty, and the new missions being slated for the agency—such as verifying a Fissile Material Cutoff Treaty and the disposal of excess material from U.S. and Russian warheads—it is not clear how the regime can live up to expectations. Perhaps most important, the political support from the United States and other key countries required to sustain the regime is not evident.

To conclude the discussion on the link between nuclear power and nuclear weapons, Paul Leventhal presents a scorecard developed by the Nuclear Control Institute on the performance of U.S. nonproliferation laws and policy. He also assesses the challenges the regime faces.

12

Nuclear Power and Nuclear Weapons in India, Pakistan, and Iran

George Perkovich

In evaluating the effectiveness of the technical and legal barriers established to maintain the separation between civil and military applications of nuclear technology, it is useful to look at actual experience. This chapter looks backward to assess how effective those barriers have been in three countries in southwest Asia: India, Pakistan, and Iran.[1]

The United States, Canada, France, and other countries cooperated with the ostensibly civil nuclear programs of India, Pakistan, and Iran before, during, and after the nuclear nonproliferation regime was being established on the foundation of the 1968 Nuclear Non-Proliferation Treaty. During this evolutionary period, the American relationship with these states and their nuclear programs shifted from positive to negative. In the beginning, the United States promoted development of nuclear technology; in the mid-1970s, it tried to constrict nuclear cooperation in an effort to forestall proliferation of nuclear weapons. These changes in approach make it difficult to assess whether any alternative mixture of cooperation and punishment would have been more effective in stemming India's and Pakistan's acquisition of nuclear weapons. Yet this history does shed light on the ways in which civil nuclear programs can be used to acquire nuclear weapons. Whatever one thinks about the relative effectiveness of punitive sanctions versus positive inducements in motivating states to abjure nuclear weapons, the international community has a clear interest in identifying pathways by which states might acquire these weapons.

India provides a rich historical case study of the connections between civilian and military applications of nuclear technology and know-how. This chapter describes briefly some of these connections. Iran provides an intriguing current and future case study. Will it become another India? The answer may be no because it is a member of the Nuclear Non-Proliferation Treaty. On the other hand, does its membership really matter? Will the regime hold? Or will civil nuclear technology contribute to a nuclear weapons program in Iran? Do the cracks in the full-scope safeguards policy and the Nuclear Suppliers Group Guidelines in any way contribute to the willingness of Russia and perhaps other countries to provide sensitive nuclear technology to Iran, and does it matter?

The following discussion of India, Iran, and, to a lesser extent, Pakistan covers some ground that has already been plowed but also adds a few furrows in answering, at least partially, the question of whether Iran is another India. The question is a very interesting one. Important differences between India and Iran suggest that the answer is no. Yet there are also similarities between the aspirations of the two countries and the modalities of developing nuclear capabilities that make this question worth exploring.

Four Themes

Deductively, it is possible to identify four themes or issues relevant to linkages between civil and military nuclear programs. First is the problem of the potential diversion of fissile materials and know-how from civilian nuclear power programs to military applications. Second is the likelihood that civilian nuclear programs train people in ways that are useful in military applications. This risk includes technical training in reactor design, plutonium separation, and metallurgy, and also in the art of procuring sensitive equipment from foreign sources. The third theme is the general political and technical cover that a civil nuclear program can provide for weapon work. Under the umbrella of legitimate civilian work, a nuclear establishment might also get away with weapon applications. Finally, a nuclear power program can create a bureaucracy, an establishment within a state, that can then affect the politics and decision making within that state as it contemplates nuclear weapons. The fourth theme, although the least remarked upon, is in fact very important. It is not a technical issue, but rather a political, institutional phenomenon that results in powerful constituencies that may favor the development of nuclear weapons and that can muster the resources to push political leaders to do so.

India was, or is, a case where all these themes came together. Its nuclear weapons program emerged from and was augmented and covered by its civil nuclear program. In the process a powerful nuclear state within a state formed that had a major impact on national decision making. Some of these links are also apparent in Iran and Pakistan.

It is worth noting here at the outset that nuclear weapon proliferation in India occurred in a very different historical context than that of today, an important point. India's nascent interest in acquiring a nuclear weapon capability, and the initial development of that capability, occurred before the Nuclear Non-Proliferation Treaty was negotiated. The United States, Canada, and other states assisted India in developing its early nuclear capability, sometimes against warnings by government experts that weapon proliferation could result. Moreover, when the treaty finally was negotiated, India chose, like Israel and Pakistan, not to sign it. These three states chose not to relinquish their juridical rights to develop nuclear weapons. Some of the considerations now being raised do not apply to India historically as they do to Iran, because Iran is a party to the treaty and has pledged not to acquire nuclear weapons.

Diversion of Nuclear Materials

That said, it is important to note that when India used the spent fuel from the Cirus nuclear reactor for the explosive that it tested in 1974, it did violate the letter and spirit of cooperation agreements it had with Canada and the United States. Some of India's plutonium stockpile today may derive from use of this reactor and related heavy water in contravention of India's pledges to these countries. If this fact leads to blaming India, then the blame should be shared, for governments and nuclear industry leaders in the United States, Canada, France, and other states were culpable "enablers" in light of the way they promoted nuclear power in India.

The United States and Canada had plenty of clues that India might "divert" its "peaceful" nuclear assistance for explosive purposes. In 1956, for example, the statute for the International Atomic Energy Agency was being negotiated, and Homi Bhabha, the chairman of the Indian Atomic Energy Commission, represented his government in the negotiations. Bhabha was the real driving force in India's nuclear weapons program. He steadfastly and effectively sought to weaken the controls that would be placed on plutonium and other special nuclear materials. He very consciously kept open India's options to divert materials, and, more broadly, the infrastructure of reactors and plutonium separation facilities, to weapon applications. In a speech during the negotiations in that year he said that the revised International Atomic Energy Agency statute ensured that fissionable material "produced in agency aided projects in a country should be at the disposal of that country, which should have the right to decide whether it wished to go ahead with a particular use of that fissionable material or not."[2] He was arguing for a reduction in the restrictions on plutonium. He went on to say, "there are many states, technically advanced, which may undertake with Agency aid, fulfilling all the present safeguards, but in addition run their own parallel programs independently of the Agency in which they could use the experience and know how obtained in Agency-aided projects, without being subject in any way to the systems of safeguards." Bhabha made those comments in 1956, and in the decades thereafter India followed essentially that same strategy.

The point is that there is a tendency to say proliferators are "bad guys" and to include India in that category. In many cases, however, proliferation depends also on suppliers, of which there were plenty willing to supply India with the wherewithal to bring its strategy about.

The Important Role of Training

The second theme is training. The key point to be made here is that training is actually as important as the diversion of fissile materials. The United States, for example, trained more than a thousand Indian atomic scientists and engineers before 1972. All of them worked in India's massive civil nuclear program, but some of them also worked on weapons. There is another example: Two of the key figures in India's effort to design and produce a neutron initiator for its first nuclear device learned much of what they needed in French laboratories, where they had been sent explicitly to gain this knowledge (a neutron initiator supplies a burst at the critical microsecond necessary to trigger a chain reaction). Developing neutron initiators is extremely difficult. Without a neutron initiator, implosion weapons will not work. India's capacity to develop neutron initiators benefited enormously from those individuals' training in France, as has Pakistan's from cooperation with China.

The Indian case also shows how training ties directly into the broader question of cover for nuclear weapon work. All the key figures in India's nuclear weapon endeavors have also worked simultaneously on civil projects. They were dual-use people. There is a tendency to talk about dual-use technology, but dual-use scientists and technologists are even more important. Civil nuclear programs enable the training of dual-use talent.

A conversation the author had with the former leader of Pakistan's nuclear program, Munir Ahmed Kahn, who died last year, drove this point home. Kahn was a delightful, urbane man and a patriotic Pakistani and citizen of the world. At that time Pakistan was trying to acquire a reprocessing plant, a red flag that dared the world to conclude that Pakistan was trying to build a bomb. The author asked Kahn why Pakistan in the 1970s, under his direction, pursued French reprocessing plants and assistance for this large nuclear infrastructure, including a reprocessing plant in Pakistan, when everyone knew that

Pakistan did not have any potential civil use for plutonium. Kahn replied,

> No, George, you don't understand. . . . The Pakistani higher education system is so poor, I have no place from which to draw talented scientists and engineers to work in our nuclear establishment. We don't have a training system for the kind of cadres we need. But, if we can get France or somebody else to come and create a broad nuclear infrastructure, and build these plants and these laboratories, I will train hundreds of my people in ways that otherwise they would never be able to be trained. And with that training, and with the blueprints and the other things that we'd get along the way, then we could set up separate plants that would not be under safeguards, that would not be built with direct foreign assistance, but I would now have the people who could do that. If I don't get the cooperation, I can't train the people to run a weapons program.

This response and its implications are a strong reason for focusing on the case of Iran and its Bushehr reactor. That reactor and the potential of spent fuel are much less an issue than the question of training Iranian cadres to design and operate a broad range of components necessary for the infrastructure to build nuclear weapons. Equally as important, as Iran seeks and works with foreign suppliers of components for this reactor, it develops Iranians with knowledge of procurement networks that they can then exploit to buy unsafeguarded components.

A Cover for Weapon Work

The third theme of the linkage between civil and military nuclear programs relates to the political or intelligence cover that a civil nuclear power program can provide for weapon work. This concern is rather obvious: The greater the number of nuclear technologies, facilities, and cadres a state has, the easier it is for it to hide weapon work. Legitimate civilian activities shield harder-to-detect military activities.

Here, again, Iran is a concern. Under the Nuclear Non-Proliferation Treaty and its Article IV, Iran has the right to develop nuclear power with assistance from other states and the International Atomic

Energy Agency, subject to safeguards. The risk that Iran's civil program can facilitate and provide a cover for weapons applications grows in direct proportion to the scale and range of technologies, facilities, and personnel encompassed by the program. The highest profile Iranian effort is the Bushehr nuclear power plant, which was unfinished at the time of the 1979 Iranian revolution and which Russia is now helping to complete. The United States has led a campaign to persuade Iran to abandon this effort and Russia not to assist it. The nonproliferation initiative is understandable but relatively less likely to succeed given that the plant itself is legitimate.

A more effective approach would be to concentrate on limiting Iran's nuclear program to this plant and getting Russia and other potential supplier states to agree not to facilitate Iran's nuclear ambitions beyond it. It might be more productive to concentrate on getting Iran to accept strictly enforceable agreements that it not build or require reprocessing, enrichment, and heavy water production capabilities, as these technologies and facilities pose the greatest risks of diversion for weapon purposes and training of cadres in weapon-useful applications. If Iran acquired facilities for these applications, they would offer a much broader cover for weapon work and exacerbate the challenge of detecting violations of the Nuclear Non-Proliferation Treaty. Thus, it might be worth exploring a "deal" with Iran (and Russia) that allowed the completion of the Bushehr power reactor in return for agreement that Iran would not acquire, on its soil, capabilities for plutonium reprocessing, uranium enrichment, and heavy water production.

The Powerful Influence of Nuclear Establishments

The fourth and last problematic link between civil and nuclear programs is political and bureaucratic. As noted, this problem deserves much more attention than it gets. Most states of proliferation concern today—Pakistan, Iran, Iraq, and North Korea—are not in the forefront of advanced technology. In each case the states are beleaguered at home, with leaderships that have ambitions of grandeur and populations that are frustrated by their countries' failures to achieve technological modernity. Iran and Iraq, like India, also represent great civilizations with rich histories. They are states that aspire to greatness on a regional, if not global, level.

Some in the United States may see nuclear technology as a fifty-year-old technology that is no longer on the cutting edge of modernity. That belief is not, or has not been, shared in India and Iran. In a country such as India—although it is less true today thanks to the information technology industry—and also in Iran, mastering the atom is seen as a manifestation of national brilliance and prowess. Once nuclear establishments are formed in these states, and once they build expensive, complicated nuclear plants and infrastructure, the "wizards" responsible for building these totems to modernity achieve a sort of mystique and become heroes. Because of this heroic status, nuclear establishments are able to exert or leverage some of their mystique and standing within their countries to influence state policy. Nuclear establishments can be seen as avatars of modernity, national prowess, and power, and the leaders of these establishments are well-positioned to persuade leaders and publics to give them rein to bring greatness to their nations.

The book, *India's Nuclear Bomb*, describes this phenomenon in India. Experience in Russia and to some extent the United States also reveals the great influence that nuclear establishments exert over national policies. The point is that to the extent that civil nuclear programs invest nuclear establishments with prestige and influence, those establishments can move their states along a path toward weapon work, and it becomes harder for political leaders to limit the establishments.

Pakistan went through an interesting experience in March 2001 that reflected the dangers of this phenomenon. Fortunately, it also showed that the government recognized the need to control its nuclear establishment more tightly. In an effort to control and consolidate this establishment so that two of its charismatic leaders could not on their own exacerbate Indo-Pakistani nuclear instability, or perhaps even establish dangerous connections outside Pakistan, General Pervez Musharraf and his government elevated the two men to other positions. One, A. Q. Kahn, the self-styled father of the Pakistani bomb, heard through a television broadcast that he was no longer running his laboratory and had been made a minister in the government. When reporters found him, he said he would not accept the position. The government persisted through the ensuing controversy, which arose precisely because Khan had become a national hero. Although the episode had a salutary ending, it illustrates well the general problem of the political power that nuclear establishments can acquire. It is very difficult to imagine a civilian government in Pakistan taking the same step.

Avoiding the Same Results

When looking to the future, there is good news on which to build. Iran's nuclear establishment has not achieved the kind of mystique and power that the nuclear avatars in India and Pakistan achieved. It is therefore very important that the United States and other nations do nothing to elevate the Iranian nuclear establishment's standing. They need to be very careful not to deny Iran, publicly and imperiously, all of its nuclear options without offering attractive alternatives. If nuclear power becomes a symbol of Iranian nationalism, in conflict with the United States and other "imperial" forces, it will be harder over the longer term for Iranian leaders to limit the nuclear program.

A final extrapolation from the Indian experience is in order. Over time, India's nuclear establishment came to be seen as a symbol of the country's modernity. Nevertheless, partly as a result of its own shortcomings, and partly because international cooperation was denied to India pursuant to the nonproliferation regime and various American laws, India's civil program languished in the 1980s and 1990s. It fell badly short of meeting every one of its targets and claims regarding the provision of electricity. At the same time, because the nuclear establishment was fairly powerful and represented so much in terms of national prowess and achievement, neither that establishment nor India's politicians would allow the program to wither.

As a result, by the 1990s weapon work emerged as the avenue where the nuclear establishment could accomplish much more than it had on the failing civil side. People started to say that they could always turn more directly to weapon work. The added attraction of weapon work was that it was inherently secretive, so that the nuclear establishment could escape the increasing accountability demanded of it on the civilian side. Politicians, bureaucrats, and journalists were beginning to complain that the nuclear establishment was not delivering electricity and was consuming lots of public funds. They began to call for audits and examinations of expenses and performance. This kind of scrutiny was less likely on the weapon side, a fact that made that work more attractive to the nuclear establishment and its champions.

Reverse Military Conversion

The literature tends not to recognize the potential of what can be called "reverse military conversion." It is, however, a very important phenomenon that must be considered when thinking through the possible links between civil and military programs.

Normal conversion is the process of switching military plants and enterprises from weapon work to civilian applications. Reverse conversion involves a shift in nuclear programs from civilian to military applications. In the future such a dynamic could have implications in Iran, Japan, and South Korea, to name the most likely states. They all have highly developed, powerful nuclear programs and live in very tough security environments. If, down the road, it appears these establishments are not going to have outlets for their creativity, are not going to get big budgets, are not going to get rewarded for civilian work over time because of doubts about the market attractiveness of nuclear power, what will they do? It must be remembered that these states live in very dangerous, unstable neighborhoods with nuclear-weapon powers around them. Will it not be tempting for their nuclear establishments and their champions to urge conversion to weapon work? Given the enormous sunk costs in training nuclear cadres and procuring elaborate facilities, will not politicians be tempted to convert these assets rather than let them languish?

It seems that no analysts or governments have addressed this problem, although it is another facet of the linkage between civil and military nuclear programs that should be addressed. Just as policy and industrial leaders in the 1950s and 1960s underestimated or overlooked the proliferation risks and effects of nuclear power programs, policymakers and opinion-shapers today must not underestimate the human complexity of reversing the connection and downsizing and constraining nuclear establishments.

If the earlier history of efforts to address the threat of proliferation is any guide, independent organizations such as the Nuclear Control Institute will be invaluable in defining the problem, identifying solutions, and motivating governments to necessary action.

13

The Nonproliferation Regime and Fissile Materials

Lawrence Scheinman

Control of access to fissile materials is the single most important roadblock to nuclear weapons. Weapon-grade material is the preferred but not the only source of material for nuclear weapons. While power reactor fuel is less efficient and perhaps more problematic, it can be used to fabricate a nuclear explosive device—for plutonium-based weapons or explosive devices, all plutonium except that containing 80 percent of the 238 isotope is usable. As such, all sources of plutonium must be regarded as a risk factor in proliferation.

Dealing with this challenge requires technical or institutional means or, more appropriately, a combination of the two. A more radical alternative would be to forswear using nuclear technology as a source of energy for social purposes. For at least three reasons that option is not viable: (1) the breadth and depth of the commitment that a number of countries have already made to peaceful nuclear energy production; (2) the continued uncertainties surrounding energy security, especially in countries substantially dependent on external sources of supply; and (3) growing concerns about the environmental impact of burning carbon-based fuels, especially coal, and continued interest in the role that nuclear-based electricity (among other alternatives) can play in addressing that problem. Nevertheless, in addition to the risk of proliferation, nuclear energy poses formidable problems relating to, among other things, environmental integrity, safety, and waste management.

Much has been said about the technical approaches to reconciling peaceful uses of nuclear technology with the risk of proliferation. Here the focus is on the international nonproliferation regime and whether it does or can provide an effective barrier against misuse of civilian fissile materials for weapons. Two assumptions underlying the discussion are that neither institutional nor technical strategies alone are sufficient to deal with the proliferation risk associated with nuclear technology, and both together offer considerably more promise than either does alone. However, to be successful even in tandem they need:

1. an unwavering *political commitment* by governments to prevent proliferation and to address it where it occurs;
2. establishment of *nonproliferation as a priority* in the political and security agendas of the governments of the leading states in the international system;
3. *leadership* by those states in strengthening the nonproliferation regime;
4. a *determined collective response* in cases of noncompliance that leaves the leaders of those states that are contemplating proliferation with no doubt that the five permanent members of the UN Security Council will stand together in dealing with such incidents.

That said, it is also true that there are *no panaceas* for unequivocally foreclosing the potential risk of proliferation. Even if nuclear energy as a means of producing electricity were eliminated, the risk would not end. States have in the past, and may well in the future, pursue the acquisition of nuclear weapons even when they have no peaceful nuclear power program from which to obtain some of the ingredients of a nuclear weapon program or by which to mask the existence of a clandestine one. Eliminating nuclear technology as a resource for energy production does not eliminate the problem of proliferation. Nor does the prospect of global nuclear disarmament and the immobilization and disposition (one way or another) of all attendant fissile materials—likely prerequisites for consigning nuclear proliferation to the dustbin of history—appear to be a near-term possibility. Furthermore, unlike the situation that prevailed from the dawn of the nuclear age through the beginning of the last decade, an even more pernicious problem has emerged. It is the prospect that what has been called the "residue" of the cold war—technology, expertise, and special nuclear materials—may end up in the wrong hands because of insufficient protection, control, and accounting for materials in the former Soviet Union and the temptation for illicit trafficking of such materials and expertise, particularly in the face of economic deterioration. The Clinton-Putin summit agreement in June 2000 called for the disposition of 68 metric tons of weapon-grade plutonium deemed no longer necessary for defense purposes, subject to verification arrangements to be concluded with the International Atomic Energy Agency. Although this agreement is a step in the direction of gaining control over this risk, there is still a long way to go.

Institutional approaches relate primarily to the nuclear nonproliferation regime, which is anchored on the Nuclear Non-Proliferation Treaty and also includes regional nuclear weapon-free zones, export controls, verification safeguards, and security assurances, as well as national legislative and administrative implementing measures. Regimes are institutional arrangements intended to bring about collective outcomes in international society through the establishment of norms, rules, and procedures that regulate and constrain behavior in a given area of activity. Participation in the treaty-based nuclear nonproliferation regime entails commitments with respect to behavior in the nuclear arena. Included are undertakings by non-nuclear-weapon states that are parties to the regime not to: (1) receive, manufacture, or otherwise acquire nuclear weapons or other nuclear explosive devices; (2) seek or receive any assistance in the manufacture of the same; and (3) submit all nuclear material in all peaceful nuclear activities to a comprehensive safeguards system to verify that they are not diverting any nuclear material from peaceful nuclear activity to nuclear weapons or other nuclear explosive devices.

Views on the relevance of regimes to the behavior of nations and their effectiveness in channeling or governing behavior vary. Hard-core realists contend that regimes are epiphenomena, embedded in a larger order of sovereign states whose structure, rules, and procedures predominate all else—for example, deterrence is a stronger explanation for state behavior than the rules of regimes. Liberal institutionalists do not argue the irrelevance of the sovereign state but do contend that contractually or consensually created regimes can and do provide frameworks within which states—big and small, strong and weak—experience benefits and learn to refine or sometimes redefine national interests in a manner consistent with the rules and principles of the regimes.

Safeguards Regimes Do Matter

Experience shows that regimes do in fact matter:

1. They *erect legal barriers* against the further spread of nuclear weapons.
2. They *embody the norm* of nonproliferation, make political decisions to acquire nuclear weapons more difficult, and raise the stakes of doing so.
3. They *serve as confidence-building measures* that help to reinforce national security.
4. They *provide a framework* within which export control, verification, and collective response measures can be formulated and implemented.

While regimes alone are not sufficient to prevent the spread of nuclear weapons, they are a critically important element in the effort to achieve that result.

Additional factors that are relevant to the credibility of the regime are delegitimization and devaluation, which involve changes by nuclear-weapon states in strategic doctrine and not just numerical reductions, reliable positive and negative security guarantees, and an effective co-operative or collective security system in which any nonaggressor state could have confidence.

The revelations regarding Iraq's extensive clandestine nuclear weapons program that emerged during inspections in the aftermath of the 1991 Gulf War underscored the shortcomings in the International Atomic Energy Agency safeguards system, in particular the need to strengthen its capacity to detect undeclared activity. Shortly afterward, the board of governors of the agency took a number of steps deemed to be within the legal authority conferred by existing full-scope safeguards agreements, including requiring:

1. additional information from states regarding facilities that have or will contain nuclear material subject to safeguards;
2. expanded use of unannounced routine inspections;
3. the collection of environmental samples at sites where inspectors already have access; and
4. confirmation of the agency's right to carry out special inspections anywhere in a state that has accepted comprehensive safeguards, if inspections are required to confirm that all nuclear material that should be under safeguards has been reported to the agency.

In addition, the agency's board of governors

1. confirmed the right of the agency to receive national intelligence reports that could facilitate its carrying out its mission; and
2. called upon all states to provide more comprehensive reporting on exports, imports, and production of nuclear material beyond that required by existing safeguards agreements.

In 1997 the board of governors endorsed the Additional Protocol to existing safeguards agreements that addressed further critical requirements for a more effective safeguards system, including:

1. information about, and access by inspectors to, all aspects of a state's nuclear fuel cycle, from mines to waste storage;
2. information about and access to all buildings on a nuclear site;
3. information about and access to all research and development related to fuel cycles;

4. information about the manufacture and export of sensitive nuclear-related technologies;
5. the ability of the inspectorate to collect environmental samples beyond declared locations when the International Atomic Energy Agency deemed necessary.

Collectively, these and the earlier safeguards measures that clarified and reinforced existing authority, together with the baseline safeguards system, offer a very comprehensive picture of a state's resources, activities, and capabilities. They are short of the perfection that might exist if direct verification of all aspects of a state's declaration could be achieved (cost is a factor here). However, they do provide the basis for having sufficient information and access to understand, dissect, and draw inferences regarding the whole of a state's nuclear program and activities and thereby to plan an effective verification strategy for the state. Another Iraqi surprise could still occur, but the measures substantially raise the bar that a state subject to comprehensive safeguards would have to scale to carry out a clandestine nuclear program successfully.

Two Significant Issues

Two significant issues lie ahead, however: (1) the financing of the safeguards system, and (2) the number and importance of the states signing the additional protocol and, even more, bringing it into force.

Zero Real Growth

It could and would be a pyrrhic victory to have a strengthened safeguards system that lacks sufficient resources to implement it as intended. Yet the International Atomic Energy Agency faces that situation today. In a statement before the board of governors, Director General Mohamed ElBaradei warned of the risk that safeguards might experience a failure given the disparity between obligations and resources. The agency has labored under a policy of zero real growth since the mid-1980s, whereas the number of states, facilities, and locations has more than doubled and the amount of nuclear material to be safeguarded has increased correspondingly. The number of significant quantities of nuclear material to be safeguarded has doubled in the past six years alone. As much as the safeguards system needed strengthening, so also does the attitude of states relying upon it, if the International Atomic Energy Agency is to break out of zero real growth.

Implementing a New Protocol

The safeguards experience in relation to Iraq exposed the limitations and weaknesses of the system and highlighted the cultural and implementation attributes needing correction. It also had the very positive effect of bringing about a strengthened safeguards system. In so doing it offers some further insight into dealing with the continuing challenge to nonproliferation posed by fissile nuclear material. If the world cannot wish fissile material away, it must deal with what is and think in terms of *how* to (1) *minimize* the presence of separated plutonium or highly enriched uranium, including through the pursuit of technical options that *eliminate* the separation of plutonium from spent fuel; (2) *control* the separated plutonium effectively; (3) progressively *reduce* the amount of separated plutonium; (4) *pace* any further reprocessing of civilian spent fuel to match the de facto requirements for insertion into reactors so that stockpiles do not build up; and (5) find ways to *secure* plutonium in spent fuel pending ultimate disposition.

This approach implies or suggests *looking for ways to build out from existing regime structures* and exploring the fashioning of institutional arrangements that can accommodate these concerns. Looking ahead requires looking back—in particular, it is important to revisit the concepts applied to dealing with plutonium and spent fuel two decades ago in the wake of the International Nuclear Fuel Cycle Evaluation. The administration of President Jimmy Carter, the proponent of the evaluation, hoped it would demonstrate that the once-through fuel cycle was the optimal choice for capturing the benefits of peaceful nuclear energy while substantially reducing the risk of proliferation. While not proving the case, the International Nuclear Fuel Cycle Evaluation did recommend exploring possibilities for international arrangements to manage separated plutonium (International Plutonium

Storage) and spent nuclear fuel (International Spent Fuel Storage). Studies were undertaken in both cases, but no conclusion was reached.

The issue of alternative means by which to address stockpiles of both military and civilian separated plutonium has re-emerged on the agenda of the International Atomic Energy Agency and other venues. One option being mentioned is exploration of accords for reporting on stocks, transfers, and storage. Non-nuclear-weapon states with substantial stocks of separated plutonium, in particular Japan, have over time become increasingly sensitive to how national stockpiling of such material is perceived by others and whether institutional arrangements might be crafted that would serve to enhance security and ameliorate concerns about the presence of these stocks. This concern would seem to open the door of opportunity to revisiting *multilateral or international plutonium and spent fuel storage* or alternative approaches and to establishing incremental institutions to complement and fortify the nonproliferation regime.

The nonproliferation regime is not static; it is a dynamic system amenable to growth and innovation. Together with continued efforts to evaluate the current fuel cycle and identify economically and technically acceptable ways to have access to nuclear technology for peaceful purposes and energy production without the proliferation risks posed by current fuel cycle choices (that is, develop systems that foreclose separation of weapon-usable material), innovative approaches may help provide a firmer grasp on the challenges posed by the presence of fissile material.

It can and will be argued that, even if progress is made along the above lines, there still remains a risk that clandestine facilities will be built to siphon off spent fuel and access the contained plutonium. Here the quality of the strengthened safeguards system comes into play, because it provides a more dependable verification capability with knowledge of who and where to target in the event of nonparticipation.

Still another concept that has been delayed in negotiation is a fissile material cutoff treaty that would foreclose any further production of fissile material for weapons or explosive purposes. This building block is likely necessary for progress in dealing with existing stockpiles resulting from military or civilian production. The 1995 Non-Proliferation

Treaty Extension Conference singled out the fissile material cutoff treaty as the next step to take following completion of a comprehensive test ban treaty. However, Chinese efforts to link any progress on a cutoff to negotiation of a treaty to prevent an arms race in outer space—a condition prompted by U.S. plans to pursue a national missile defense—has blocked progress on this issue in the Conference on Disarmament.

Another important factor is a third dimension of the lessons that came out of the Iraqi experience. As then-Director General Hans Blix asserted shortly after the revelations of clandestine activity, not only is access to information and to sites necessary if the agency is to provide a high degree of assurance that clandestine activities can be discovered, but so is assurance that "access to the Security Council is available for backing and support that may be necessary to perform the inspection." The Security Council gave that assurance in 1992 when British Prime Minister John Major, speaking on its behalf, asserted that proliferation of any kind constituted a threat to international peace and security. This assertion raises the prospect of opening Chapter VII of the UN Charter and allowing for a range of responsive measures, including force. Major further stated that the council would take appropriate measures when the International Atomic Energy Agency notified it of safeguards violations. There has been one notification involving North Korea, but the Security Council was unable to agree on how to deal with the problem and left it to the United States to seek a solution. In that sense the Security Council has yet to deliver on its commitments. If and when it does stand unified and resolute on proliferation and violations of treaty undertakings in the nuclear field, the uncertainties surrounding nuclear power and fissile material will diminish, the presence of the atom will become that much less threatening, and its use for benign social and economic purposes will be more manageable.

While determined collective response to the threat of proliferation is critically important, there is a need to move forward on other fronts as well, with a focus on developing and strengthening institutions that help avert the threats of proliferation in the first instance. Such institutions require measures that both address the incentives to proliferate and curtail or foreclose opportunities to do so. Here the focus in on the latter. The tendency toward system inertia in the absence of a sense of impending crisis underscores the need to call forth three elements identified earlier as essential to managing the atom: political will and commitment; prioritization as a matter of national policy; and leadership in the international arena.

These elements should be harnessed to an agenda consisting of at least the following three things. The first is to implement the Additional Protocol to State Safeguards Agreements (also called the 93 + 2 Additional Protocol) fully, so as to increase the opportunity for the International Atomic Energy Agency to carry out comprehensive safeguards effectively in all of the non-nuclear-weapon states party to the Nuclear Non-Proliferation Treaty. To the extent that nuclear-weapon states extend their existing voluntary agreements to the strengthened safeguards regime, the agency would apply them there as well. The weapon states could ill afford to demur on such acceptance if they wish to see the strengthened system applied in all of the non-nuclear-weapon states that are parties to the Nuclear Non-Proliferation Treaty—any sense of discrimination would work against the interest of strengthened safeguards.

The second is to negotiate a fissile material cutoff treaty that obliges nations to end the production of fissile materials for military or explosive purposes and to bring all fissile materials under verification. The UN General Assembly adopted a consensus resolution supporting a cutoff treaty in 1993, and in 1995 the decision of the Non-Proliferation Treaty Review and Extension Conference on Principles and Objectives identified fissile cutoff as a priority goal for "immediate commencement and early conclusion." The Conference on Disarmament, the negotiating forum for a fissile material cutoff treaty, needs to end the stalemate that has precluded implementation of this consensus objective for the past four years.

Third, given the presence of separated plutonium in a number of locations around the world and the attendant uncertainty and suspicion it generates, there should be reconsideration of earlier concepts of plutonium storage under terms and conditions that would diminish if not eliminate the concerns of the international community with the presence of such weapon-usable material.

Progress along these lines would close important gaps in the current nonproliferation system and serve as additive building blocks that contribute to reconciling efforts to enable nuclear power to play a constructive role in meeting global energy needs without at the same time increasing the risk of proliferation. Fully achieving that objective will, as stated, require the full effort of all states to move toward security in a nuclear weapon–free world.

Closing Thoughts on Nonproliferation: The Need for Rigor

Paul L. Leventhal

Adrian "Butch" Fisher, the first general counsel of the Arms Control and Disarmament Agency and the chief U.S. negotiator of the Nuclear Non-Proliferation Treaty, once testified on Capitol Hill that the treaty does not require the United States "to do anything foolish."[1] He was referring to two provisions. One, in Article III, permits nuclear exports to nations that refuse to join the treaty so long as they accept international inspections on the exported items, albeit not on their entire programs. The second is the obligation under Article IV to supply nations that do belong to the treaty and do accept universal inspections. Fisher's point was a simple one, what he called "a truism":[2] there is nothing in the treaty that requires the United States to supply parties or nonparties to the treaty with anything that does not make economic sense because, "If it does not make sense economically, a reasonable person would say you really have a weapons capability in the back of your mind."[3]

Fisher's wise words of more than a quarter century ago come to mind when considering today's dangerous state of affairs, as illuminated by Davis, Perkovich, and Scheinman. Davis warns of cracks developing in the nonproliferation regime. Perkovich points to the weapons developed by India and Pakistan under cover of their "peaceful" nuclear imports, and the possibility that Iran is embarked on the same course. Scheinman describes a difficult search for international measures to ensure that all of the world's excess civilian plutonium can be kept under control. All of these troubles have a common cause: the failure of the United States to take the opportunity it had when it was still calling the shots in the 1960s and 1970s as the result of its virtual monopoly on nuclear reactors and enriched nuclear fuel. It did not effectively pursue its supreme national interest in preventing the spread of nuclear weapons.

What turned wise men like Fisher and former Atomic Energy Commission chairman David Lilienthal sour on "Atoms for Peace" in the mid-1970s, when the Senate Government Operations Committee launched an extensive, ground-breaking set of hearings on nuclear nonproliferation, was clear evidence that the United States was acting foolishly. Indeed, Lilienthal's memorable words at those hearings were even stronger than Fisher's. He said: "If we adopt [a] defensive, apologetic view of America's posture respecting the further spread of nuclear weapons, you can be sure that once more we will be taken advantage of, because of our national inclination to want to be loved in foreign parts, even at the cost of being the atomic patsy of the world—which is what we have become."[4] He specifically had in mind what he called the "morally indefensible doctrine that if our manufacturers and vendors do not continue to supply these potentially deadly materials and this technology, the manufacturers of other countries will do so."[5]

Perkovich is on the mark when he states that the blame for India's misuse of peaceful plutonium for its original test of 1974 and for building its now declared arsenal is not India's alone. It must be shared by India's original suppliers, the United States and Canada, and by other "culpable 'enablers'" along the way, including France and Russia.

Today's world has been made safe *for* plutonium, not *from* plutonium. The thanks for that go to a nonproliferation regime dominated by industrial and bureaucratic interests that narrowly interpret the letter of the treaty while ignoring its spirit and the early admonitions of

Fisher and Lilienthal. Scheinman alludes to a basic bargain implicit in the treaty. Article IV states that in return for adherence to the treaty, states without nuclear weapons are entitled to full access to peaceful nuclear technology. Article VI requires that the nuclear-weapon states pursue disarmament in return for non-nuclear-weapon states abjuring nuclear arms.

There is, however, still another bargain implicit in the treaty, one that is rarely discussed. The Nuclear Control Institute has referred to it as a "dynamic tension" between Articles I and II, on the one hand, and Article IV on the other. The treaty requires that a transfer of nuclear technology authorized by Article IV must be "in conformity with" the solemn obligation of all parties under Articles I and II not in any way to transfer nuclear weapons capability to non-nuclear-weapon states. The Nuclear Control Institute's counsel, Eldon Greenberg, did an excellent legal analysis of this issue in the runup to the 1995 Nuclear Non-Proliferation Treaty Extension Conference. He made the point that if there is no economic justification for the use of plutonium as fuel—surely the current situation, given the demise of the breeder reactor and the severe diseconomics of plutonium in conventional reactors—then the treaty, as it now exists and without any need to amend it, can be interpreted to say that any production of separated, weapon-capable plutonium is a violation of Articles I and II. That is, the transfer of any technology, or the continuation of any programs, for the utilization of separated plutonium run afoul of the Article I and II prohibitions.[6] Like Fisher's earlier economic test for the legitimacy of nuclear exports, this application of an economic test for plutonium's legitimacy under the treaty also was dutifully ignored.

In assessing the Nuclear Non-Proliferation Treaty regime, it is essential to put first things first. The premier question is whether continued civilian use of plutonium and the other atom bomb material, highly enriched uranium, makes any sense in terms of the future of the nuclear industry and the security of the world. Bertram Wolfe, the former General Electric nuclear chief whose chapter appears in the final section, has acknowledged to the author over the last twenty-five years that plutonium is not in the best interest of the nuclear industry because it raises all kinds of controversies and dangers and is not an essential fuel if no breeder is immediately at hand. Yet commerce in plutonium continues unabated.

What the Nuclear Non-Proliferation Treaty regime most lacks today is a sense of rigor. It is squishy soft on the question of the production by the ton of materials that can be used by the pound to build nuclear weapons. Years ago, Harold Feiveson of Princeton University, whose chapter also appears in the last section, made an interesting point that bears on this subject—a need for "symmetry," for symmetrical obligations between weapon and nonweapon states. There is a need for symmetry and reciprocity between those with and without nuclear weapons to get rid of their excess plutonium.

Today that concept is being applied in reverse—to find ways to perpetuate the plutonium business rather than build a treaty regime to enable its shutdown. The nonweapon states, faced with electric utility companies that cannot absorb all the plutonium fuel being produced and that prefer not to use it at all, are now scrambling for a fig leaf to hide their peaceful plutonium stockpiles behind. The latest proposal, as this book is being readied to go to press, is international plutonium storage, but with a twist. Rather than establish an international repository removed from the nations that produce plutonium, as was contemplated in Article XII.A.5 of the Statute of the International Atomic Energy Agency, each plutonium-producing state would have its own "international" repository, inspected by and under the custody of the agency but operated and protected by the state, for storing the state's excess plutonium and for receiving additional excess plutonium as it is separated from spent fuel. Each state, according to this proposal, could withdraw plutonium from the repository upon request to the agency.[7] Perhaps a reality check on such a transparently self-serving plan would be to take opinion polls in the Republic of Korea (South Korea), Indonesia, China, and Taiwan to determine whether their citizens would be any less nervous about Japan's large (and still growing) stockpile of plutonium if the stockpile had an International Atomic Energy Agency flag flying over it. Meanwhile, the United States and Russia are planning to get rid of their excess weapon plutonium, not by disposing of it as waste, but by establishing their own commercial plutonium industries for turning it into fuel for nuclear power plants.

In short, there is lack of rigor in how the world deals with plutonium. We are not yet at the point that plutonium is regarded for what it is: an atom bomb material too dangerous to use in commerce. It may be that civilian plutonium or bomb-grade uranium will have to be exploded in bombs in a war or an act of terrorism before the reality of the danger is universally accepted. Increasing worries about the next war between India and Pakistan turning nuclear, and about post–September 11 terrorists using stolen fissile materials to make crude atomic bombs, may bring the world to a new level of awareness—but hopefully well short of the precipice.

For now, a nuclear industry bias still pervades all the councils addressing nonproliferation and makes dealing rigorously with the plutonium danger all but impossible. One veteran journalist who is a longtime observer of these matters once remarked to the author that the problem with the Nuclear Non-Proliferation Treaty Review Conferences is that "the nukeheads run the show." His phrasing was perhaps indelicate, but his basic point about the dominant role of the nuclear industry and nuclear bureaucracy at these meetings is sound. It may be impossible to do anything different with regard to plutonium unless these two powerful interests can be motivated to change the way they think and act.

Rudolph Rometsch, the first and only inspector general of the International Atomic Energy Agency, some years ago made the same observation to the author regarding the nuclear establishment in India that Perkovich makes in his essay. "It is like dealing with a state within a state," Rometsch remarked, and he suggested that this situation was not good for the effective application of the agency's safeguards in India on the few imported nuclear plants that accepted them. This came from an official so powerful and outspoken that upon his retirement his inspector general's title was retired with him.

As Perkovich suggests, the state-within-a-state phenomenon is endemic among countries with large nuclear programs. The phenomenon has two impacts. Perkovich noted the impact on the politics and decision making within those states, with the potential (already realized in India and Pakistan) for a shift in nuclear programs from civilian to military applications—a process he calls "reverse military conversion." The other impact is found at the international forums held under the auspices of the International Atomic Energy Agency (especially meetings of its board of governors) and the Nuclear Non-Proliferation Treaty—a powerful bias that ensures the perpetuation of the plutonium business.

There is a ray of hope. It is found in the difference in the way the international nuclear community has dealt with bomb-grade, highly enriched uranium compared with plutonium. Nearly all commerce in highly enriched uranium has ended, at least that originating from the United States, which is virtually the exclusive source of the fuel for research reactors in the western world. Only the Kingdom of Bavaria is insisting on building a new research reactor to use highly enriched uranium (an aberration that the German government promises to deal with in due course).

At the same time, however, commerce in plutonium is expanding in the face of adverse economics. Why is it that nations can agree to get rid of highly enriched uranium and not plutonium? Both are atom bomb materials, although Luis Alvarez, a scientist in the Manhattan Project, once observed that "[W]ith modern weapons-grade uranium . . . terrorists, if they had such material, would have a good chance of setting off a high-yield explosion simply by dropping one half of the material onto the other half. . . . Even a high school student could make a bomb in short order."[8] The relative ease with which highly enriched uranium could be turned into bombs gave the nuclear community pause.

An international consensus emerged in the 1970s to replace the bomb-grade uranium used in research and test reactors with a high-density, low-enriched fuel that is unsuitable for weapons. The consensus was reached in the International Nuclear Fuel Cycle Evaluation during the administration of President Jimmy Carter, and it was made possible by the absence of a strong industrial bias for retaining bomb-grade uranium such as there was, and continues to be, for retaining plutonium. The research reactor community found common ground in the challenge of developing and testing a new, alternative fuel and in the realization that continued use of bomb-grade fuel in poorly guarded academic settings was too risky. A remarkable team at Argonne National Laboratory has led an international program to get the difficult job done, and the goal is now in sight. There have been some significant political obstacles along the way, especially from operators of a few larger research reactors in Europe and the United States that balked at the costs and inconvenience of going through the process of converting to low-enriched fuel. A crisis point came during the administration of President Ronald W. Reagan, when—in response to these objections—the executive branch attempted to zero out the budget for the Reduced Enrichment for Research and Test Reactor program. The Nuclear Control Institute—with a special debt to Alan Kuperman, then its issues director and now its senior policy analyst—played a critical role in helping keep the program alive. The program was kept alive because of op-ed articles and special appeals to the few members of Congress who were willing to fight to restore the $1.2 million a year that the executive branch said it could not afford to keep the conversion program going and keep bomb-grade uranium out of the hands of terrorists. This process had to be repeated over a number of years.

With determination and tenacity, the job of getting highly enriched uranium fuel out of commerce is getting done. The question is, can the same be accomplished with plutonium? There is an opportunity, but so far a global industry and a bureaucracy—a world within a world—that cannot see beyond its own special interests has squandered the chance. The immediate aftermath of the end of the cold war presented a grand opportunity for dealing with both military and civilian plutonium. The superpowers needed to begin getting rid of excess plutonium from the warheads they were retiring, and at the same time the civilian plutonium producers in Europe and Japan needed to figure out what to do with their growing surpluses.

Was there a way to deal *symmetrically* with both problems? In 1994, at the first annual plutonium disposition meeting in Leesburg, Virginia, the Nuclear Control Institute proposed that both excess military and excess civilian plutonium be immobilized in the highly radioactive waste from which it was originally separated.[9] Immobilization, the institute suggested, could serve as a "magnet" to draw in *all* of the world's excess separated plutonium and get it out of harm's way. In this way a means could be found to establish symmetry and reciprocity between the nuclear-weapon and non-nuclear-weapon states for isolating atom bomb material and preparing it for eventual geological disposal. Until final disposal, civil and military excess plutonium would be in a vitrified (glass) form equivalent to highly radioactive spent fuel (the "spent fuel standard") and thus inaccessible for direct use in weapons. This plan was consistent with the objectives of Articles I, II, III, IV, and VI of the Nuclear Non-Proliferation Treaty. Nations could work together to remove the overburden of plutonium and make the world safer.

What was the response to this proposal? Predictably, governments and industry rejected it as a "nonstarter." The institute was advised not to waste time even thinking about it. It was an interesting idea, the institute was told, but an intellectual conceit that would go nowhere.

Since then, high costs and high risks have stalled the plan preferred by the United States and Russia—to turn excess military plutonium into fuel for power reactors. This program, which seemed destined to revitalize a moribund plutonium industry in both countries, is now taking shape as a white elephant. The administration of President George W. Bush appeared at first to have withdrawn support for both the plutonium immobilization and plutonium fuel programs, but as this volume goes to press, it has predictably thrown its support to the plutonium-fuel approach. It is doing so despite the costs, risks, and complexities that make successful implementation in the United States and Russia seem highly unlikely.

The default option for the world's excess military and civilian plutonium now appears to be stockpiling, and that is not good news for the world. The problem is that those who still think plutonium is the greatest thing since sliced bread are making the policy. Somehow policy making must be taken away from those people.

Market forces might yet get the job done. Consider the case of Japan. Given the problems facing its economy, how much more abuse will it be willing to sustain to perpetuate an uneconomic, unpopular plutonium industry? It is extremely difficult in a country like Japan, where decisions on plutonium and other major issues are made laboriously by consensus, to change a decision once made; it is like trying to turn an ocean liner on a dime. If Japan does change its mind, however, the European plutonium industry loses its number one customer. If Europe and Japan are not pushing for plutonium anymore, the U.S. government can finally agree to do for plutonium what it once did for highly enriched uranium. At that point the nonproliferation regime can finally respond with rigor. The Board of Governors of the International Atomic Energy Agency for once would be in a position to pull in the right direction rather than in the plutonium direction.

Unfortunately, this situation does not pertain today. What the world has instead is a highly discriminatory, two-tiered system in which an exclusive club reserves for itself the privilege of exploiting atom bomb material for purportedly commercial uses while denying others this right. The system is unsustainable because it is discriminatory. India and Pakistan already have crashed the club, Iraq tried, and Iran is knocking on the door. Given the tons of atom-bomb material already in commerce, terrorists could knock down the door without warning.

Essentially, the nonproliferation regime has a choice to make: End the discriminatory system either by making the technology available to all or by foreclosing the option for all. The correct choice should be obvious. The current system, whether its goal is nonproliferation or counterterrorism, is living on borrowed time and will succumb to political and technological forces that are intrinsic to the system. It is unrealistic to assume a "global norm" can be sustained indefinitely that reserves only to some the right to exploit plutonium and bomb-grade uranium for peaceful purposes.

Unfortunately, beyond market forces there may be no dynamic for positive change other than a catastrophe or a close call that forces everyone to reconsider. As Richard Garwin once put it so adroitly, "One nuclear explosion can ruin your whole day." It is to be hoped that common sense will yet intervene ahead of catastrophe.

PART IV

Three Closing Views

15

An Industrialist's View

Bertram Wolfe

In 1945 the United States dropped atomic bombs on Japan, and shortly thereafter the war ended. Until 1954 no country carried out nuclear work other than for military purposes besides the original five nuclear states. In a talk that same year President Dwight D. Eisenhower said that what bothered him about nuclear work was that some twenty countries were trying to develop the ability to make bombs. He concluded by noting that maybe all countries would end up with bombs. To get around that possibility, the president proposed a peaceful nuclear energy program: The United States would agree to provide peaceful nuclear technology to any country that agreed not to make bombs.

In 1968 the Nuclear Non-Proliferation Treaty was developed, and some one hundred and seventy countries have signed it. The outcome was a lessening of the problem of nuclear weapons. Nevertheless, today Pakistan and India have nuclear weapons, and there is concern that Iran and Iraq will achieve them. Still, Eisenhower would probably say that he helped reduce the nuclear-weapon problem.

When the United States started building nuclear plants in this country after 1954, it had a tremendous need for energy. In the late sixties and early seventies the supply of energy expanded rapidly as usage doubled every ten years. There were tremendous numbers of orders for both coal and nuclear plants. The industry was selling thirty to forty nuclear plants a year, and the expectation was that over a thousand nuclear plants would be operating by the end of the century.

Then, in 1973, the world experienced the Arab oil boycott, which caused energy prices to rise and energy growth to decline. Instead of doubling every ten years, energy growth fell to a rate of 2 percent a year, a doubling every thirty-five years. As a result, since 1973 this country has had surplus capacity.

Before 1973 the Sierra Club had been in favor of nuclear power. The Sierra Club was one of the main reasons the Diablo Canyon facility was built. After 1973 the United States did not need additional plants, and the Sierra Club and the environmental movement opposed nuclear, coal, gas, oil, hydroelectric, and geothermal developments. The Sierra Club and the environmental movement supported only solar and wind power. The fact is that from 1973 to the present day the United States has had no need for new power.

Then, suddenly, the state of California found itself in terrible trouble. It experienced blackouts. The presumption was that other parts of the country would also experience blackouts when summer came and there was a high demand for air conditioning. For the first time since 1973 the United States needs new power, and the situation is serious. Another important fact is that after 1954, when the United States started building nuclear power plants, it became the world leader, with sixty new nuclear plants built up to 1973. After 1973 orders for new nuclear plants in this country came to an end. At the present time nuclear power accounts for 20 percent of the capacity in this country. It actually produces more electricity today than the entire country used in 1954, when nuclear power just started.

The critical point is that the United States and other countries now need new power facilities. It is estimated that in the next half century the world's population will increase to 10 billion people. If each person uses a third of the energy currently consumed in the United States per capita, energy use worldwide will triple. It is hard to see any way to produce that energy without nuclear power, given the projections that oil and gas will run out this century and coal in the next. The only solution is a major worldwide expansion of nuclear power.

A further point is that this country has really not used reprocessing. France does. With light-water reactors, reprocessing is not necessary. On the other hand, the world may run out of uranium in the next fifty years, as nuclear power grows. In that eventuality the only solution is the fast reactor. As such, the Nuclear Control Institute should be looking ahead at how the United States can develop the necessary fast reactors in twenty to forty years in a way that will prevent proliferation. Argonne National Laboratory and General Electric are already working on the development of a fast-breeder reactor in which the plutonium does not get separated out of the reactor. Instead, it keeps recycling. This appears to be a way to solve the problem.

The world should be looking at ways to develop the necessary fast reactors in the next thirty to forty years while staying away from proliferation. It is always possible to wait thirty years to act, just as California waited until the recent crisis forced it to recognize that it needed more power. That approach could cause worldwide devastation in the future.

16

An Arms Controller's View

Harold A. Feiveson

When people at the U.S. Arms Control and Disarmament Agency first drafted Article 3 (the Safeguards Article) of the Nuclear Non-Proliferation Treaty,[1] the assumption was that safeguards would control diversions from nuclear power programs. Initially, most arms controllers believed that as long as nuclear power remained based on a once-through fuel cycle, without reprocessing and recycling of plutonium, it could be highly proliferation-resistant—that the real cause for concern was the onset of breeder reactors. This indeed has long been the view of the Nuclear Control Institute.

However, present prospects of a very large-scale future for nuclear power make that belief less tenable. If nuclear power is to make any substantial dent in the greenhouse problem, then the nuclear power system will have to grow to ten to twenty times its present size. What are the implications of a nuclear power system that size? Technical solutions can probably address the problem of nuclear waste, although a nuclear power system of that scope would generate roughly one Yucca Mountain of waste every one to two years. Probably it would also be possible to make reactors safe, even as nuclear power expands. One analogy that people have often used is the airplane industry: As the airplane industry has diffused all over the world, with every nation having its own national airline, safety has actually improved. Thus, the spread of nuclear power throughout the world does not necessarily mean that it will become less safe.

Finally, there is the matter of proliferation resistance. Here, too, many may believe that the airplane analogy applies. Why do people not worry about bomber-resistant civil air technology? It is likely that at the start of the air industry somebody said, "You know, if you have civilian air, it could always be misused for bombers." That did not happen, however, and missiles have made the question of bomber-resistant airplane technology completely moot.

The author is not, however, sanguine that new technologies or changes in the international system will likewise make issues of proliferation resistance moot. The world should be concerned about the development of a civil nuclear industry that will, whatever the circumstances, produce a tremendous legacy of materials in thirty, forty, fifty, seventy-five years that are going to have to be safeguarded forever. The analogy here is the residue of the cold war, which has left tens of thousands of nuclear weapons and tremendous stocks of fissile materials that have to be controlled and then somehow gotten rid of.

A ten to twenty times expanded nuclear power system probably will lead to breeder reactors and recycling, with all the problems these activities imply. Possibly this will not be necessary if cheap uranium from seawater allows the long-term continuation of reliance on the once-through fuel cycle. Even here, however, a back-of-the-envelope calculation gives rise to concern. Assume a capacity of 3,500 gigawatts, a tenfold increase over today. Next assume a 100-megawatt pebble-bed reactor that uses 8-percent-enriched uranium with a burnup of about 80 megawatt-days per kilogram. Producing the fuel for this reactor will require roughly 20,000 kilograms of separative work units per year per reactor. Now assume a plant with a capacity of 1 million kilograms of separative work units per year, which is about half the capacity of the Urenco plant. Since a plant of this size could service about fifty reactors a year, that size seems pretty plausible. Based on these figures, something on the order of six hundred and fifty of these 1-million-separative-work-unit uranium enrichment plants would be operating around the world. If each plant started with natural uranium, it could make three hundred bombs' worth of weapon-grade uranium per year. If each started with 8-percent-enriched uranium instead of natural uranium, each could make maybe six times that amount, or about eighteen hundred bombs per year.

The world would then have a tremendous amount of uranium being produced out of seawater and transported around the world. There would be a tremendous amount of enrichment going on, and great incentive for scientists and engineers to devise ways to make uranium enrichment cheaper, quicker, and faster.

Perhaps it is possible to safeguard that kind of system, but the prospect gives pause for thought. In the long run, nuclear power of a magnitude to address the greenhouse challenge probably would be very difficult to make proliferation-resistant even if reprocessing and recycling could be avoided. Perhaps the only way to make a tenfold to twentyfold increased nuclear system adequately proliferation-resistant would be to require that nuclear power be placed in very centralized international parks under international control. Whether this approach would work is open to question, as countries would have to give up sovereignty over their energy systems.

17

A Historian's View

William Lanouette

In 1969 there were five nuclear weapon states: the United States, the Soviet Union, the United Kingdom, France, and China. Since then, there have been dire warnings that the number would rise, and there is evidence that Israel, South Africa, India, and Pakistan have joined the club.

There is also evidence that five nations could have built a bomb. According to press accounts, Norway, Sweden, Canada, and South Korea have the capability to do so but have decided not to. If not for the apparent threat of the administration of President Gerald R. Ford to end further aid if South Korea pursued a nuclear program, it would probably still be doing so.

With a number of countries having gone nuclear since 1969, the question remains—what else is required for people to take this problem seriously? What will it take to get the public—particularly responsible policymakers—to treat the issue of nuclear proliferation in the context of nuclear power with true concern? Will it take yet another state going nuclear? Will it take a publicized theft of nuclear material? Will it take a scandal involving nuclear secrets? Mordecai Vanunu and Wen Ho Lee got a lot of ink, but did it cause people to think that nuclear weapons pose a threat to all society? Is it going to take a scare? Will global warming be the scare that causes a re-examination of nuclear power?

The Zanger Committee raised the issue with its trigger list. The press found the trigger list to be highly useful, because the list kept its keepers debating endlessly. If they made export items on the list too specific, they were in effect giving blueprints to potential proliferators. If, however, they made the items too general, they were creating loopholes through which dangerously useful technology could be exported. That sort of debate is typical of nuclear technology and the problems that policymakers have in dealing with it.

The end of the bipolar cold war should be a time to focus on nuclear power and nuclear weapons. Perhaps some truly original thinking is needed. Leo Szilard often provided highly original thinking about the control of nuclear weapons, sometimes to good benefit.[1] His truly original approach to weapons testing, for example, one that showed him to be a contrary genius, was to go against the right thinking of the cofounders of the Pugwash movement,[2] who favored the elimination of nuclear tests. Szilard disagreed, saying instead that to stop nuclear weapons, it was necessary to "test them, test them *all*."

What may be needed to get the world through the next nuclear era is some truly original—perhaps even irreverent—thinking.

APPENDIX 1

Can Terrorists Build Nuclear Weapons?

J. Carson Mark, Theodore Taylor, Eugene Eyster, William Maraman, Jacob Wechsler

General Observations

Two options for nuclear devices to be built by terrorists are considered here: That of using the earliest design principles in a so-called crude design and that of using more advanced principles in a so-called sophisticated design.

A crude design is one in which either of the methods successfully demonstrated in 1945—the gun type and the implosion type—is applied. In the gun type, a subcritical piece of fissile material (the projectile) is fired rapidly into another subcritical piece (the target) such that the final assembly is supercritical without a change in the density of the material. In the implosion type, a near-critical piece of fissile material is compressed by a converging shock wave resulting from the detonation of a surrounding layer of high explosive and becomes supercritical because of its increase in density.

A small, sophisticated design is one with a diameter of about 1 or 2 feet and a weight of one hundred to a few hundred pounds, so that it is readily transportable (for example, in the trunk of a standard car). Its size and weight may be compared with that of a crude design, which would be on the order of a ton or more and require a larger vehicle. It would also be possible, in about the same size and weight as a crude model but using a more sophisticated design, to build a device requiring a smaller amount of fissile material to achieve similar effects.

For a finished implosion device using a crude design, terrorists would need something like a critical mass of uranium (U) or plutonium (Pu) or, possibly, UO2 (uranium oxide) or Pu02 (plutonium oxide). For a gun-type device, substantially more than a critical mass of uranium is needed, and plutonium cannot be used. It may be assumed that the terrorists would have acquired (or plan to acquire) such an amount either in the form of oxide powder (such as might be found in a fuel fabrication plant), in the form of finished fuel elements for a reactor—whether power, research, or breeder—or as spent fuel.

For a small, sophisticated design, the terrorists may need a similar amount of fissile material, since practically all the presumed reductions in size and weight have to be taken from the assembly mechanism, and, with a less powerful assembly, not only will it be important to have the active material in its most effective form, but its amount will have to be sufficient to achieve supercriticality. Alternatively, a smaller amount could be used in a sophisticated design with a more powerful and heavier assembly mechanism.

Conceivably, oxide powder might be used as is, although terrorists might choose to go through the chemical operation of reducing it to metal. Such a process would take a number of days and would require specialized equipment and techniques, but these could certainly be within the reach of a dedicated technical team.

Fuel elements of any type will have to be subjected to chemical processing to separate the fissile material they may contain from the inert cladding material or other diluents. This process would also require specialized equipment, a supply of appropriate reagents, well-developed techniques specific to the materials handled, and at least a few days to conduct the operation. Spent fuel from power reactors would contain some plutonium but at such low concentrations that it would have to be separated from the other materials in the fuel. It would also contain enough radioactive fission fragments that the chemical separation process would have to be carried out by remote operation, a very complicated undertaking requiring months to set up and check out, as well as many days for the processing itself. The fresh fuel for almost all power reactors would be of no use, since the uranium enrichment is too low to provide an explosive chain reaction.

The terrorists would need something like a critical mass of the material they propose to use. For a particular fissile material, the amount that constitutes a critical mass can vary widely depending on its density, the characteristics (thickness and material) of the reflector employed, and the nature and fractional quantity of any inert diluents present (such as the oxygen in uranium oxide, the uranium 238 in partially enriched uranium 235, or chemical impurities).

For comparison purposes, it is convenient to note the critical masses with no reflector present (the "bare crit") of a few representative materials at some standard density. For this discussion, the following examples of bare critical masses have been chosen:

10 kilograms (kg) of Pu 239, alpha-phase metal (density = 19.86 grams per cubic centimeter [gm/cc]).
52 kg of 94% U-235 (6% U-238) metal (density = 18.7 gm/cc).
approximately 110 kg of U02 (94% U-235) at full crystal density (density = 1I gm/cc). approximately 35 kg of Pu02 at full crystal density (density = 11.4 gm/cc).

In all cases (others as well as these), the mass required for a bare crit varies inversely as the square of the density. Thus, the bare crit of delta-phase plutonium metal (density = 15.6 gm/cc) is about 16 kg. Similarly, at densities the square root of two times larger than those above, the bare crit masses would be one-half those indicated. If any reflector is present, the mass required to constitute a critical assembly would be smaller than those above. With a reflector several inches thick, made of any of several fairly readily available materials (such as uranium, iron, or graphite, for example), the critical mass would be about half the bare crit. Thicker reflectors would further reduce the mass but would be more awkward without providing much more of a reduction. (Although beryllium is particularly effective in this respect—providing critical masses as low as one-third the bare crit—it is not readily available in the form needed and is not considered further.) It is consequently assumed here that a mass of half the bare crit is what terrorists would require to complete a near-critical (crude) assembly.

With respect to the effects of dilution by isotopes of heavy elements, only the two most obvious cases need be considered. One is that of reactor-grade plutonium. This material is not uniquely specified, since the fractional amount of the Pu-240 depends on the level of exposure of the fuel in the reactor before it is discharged. However, at burn-up levels somewhat higher than present practice, the bare crit of plutonium would be only some 25–35 percent higher than that for pure Pu-239. Because of spontaneous fission, the effect of the Pu-240 on the neutron source in the material is thus likely to be more important than its effect on the critical mass. Nevertheless, nuclear weapons can be made with reactor-grade plutonium.

The other obvious dilution case is that of uranium at enrichments lower than 94 percent. Here the effect on critical mass, and consequently on the amount of material that must be acquired and moved by the assembly system, is quite appreciable. For example, the bare crit of 50 percent enriched uranium is about 160 kg (~3 times that of 94 percent material) and for 20 percent material about 800 kg (~15 times that for 94 percent). Similar factors will apply for uranium oxide as a function of enrichment. In this same connection, it may be noted that the mixed oxide fuel once considered for the Clinch River Breeder Reactor (~22 percent plutonium oxide plus ~78 percent uranium oxide) would correspond to uranium at an enrichment of somewhat less than 40 percent and have a critical mass a little more than four times larger than 94 percent uranium oxide.

As a final general observation, for a crude design, terrorists would need something like 5 or 6 kg of plutonium or 25 kg of very highly enriched uranium (and more for a gun-type device), even if they planned to use metal. They would have to acquire more material than is to go into the device, since with metal considerably more material is required to work with than will appear in the finished pieces. The amounts they would need can be compared with the formula quantities identified in federal regulations for the protection of nuclear materials: 5 kg U-235, or 2 kg plutonium. Sites at which more than a formula quantity is present are required to take measures to cope with a determined, violent assault by a dedicated, well-trained, and well-armed group with the ability to operate as two or more teams. Transport vehicles that carry more than a formula quantity must be accompanied by armed escort teams and have secure communications with their base. Transport vehicles carrying smaller amounts are not so heavily guarded, but there are provisions intended to ensure that in the aggregate no more than a formula quantity is on the road at one time. For terrorists having to acquire at least several formula quantities, there are formidable barriers to overcome.

Crude Designs

Crude designs are discussed primarily in the context of the problems facing a terrorist group. Schematic drawings of fission explosive devices of the earliest types showing in a qualitative way the principles used in achieving the first fission explosions are widely available. However, the detailed design drawings and specifications that are essential before it is possible to plan the fabrication of actual parts are not available. The preparation of these drawings requires a large number of man-hours and the direct participation of individuals thoroughly informed in several quite distinct areas: the physical, chemical, and metallurgical properties of the various materials to be used, as well as the characteristics affecting their fabrication; neutronic properties; radiation effects, both nuclear and biological; technology concerning high explosives and/or chemical propellants; some hydrodynamics; electrical circuitry; and others.

It is exceedingly unlikely that any single individual, even after years of assiduous preparation, could equip himself to proceed confidently in each part of this diverse range of necessary knowledge and skills, so that it may be assumed that a team would have to be involved. The number of specialists required would depend on the background and experience of those enlisted, but their number could scarcely be fewer than three or four and might well have to be more. The members of the team would have to be chosen not only on the basis of their technical knowledge, experience, and skills but also on their willingness to apply their talents to such a project, although their susceptibility to coercion or considerations of personal gain could be factors. In any event, the necessary attributes would be quite distinct from the paramilitary capability most often supposed to typify terrorists.

Assuming the existence of a subnational group equipped for the activist role of acquiring the necessary fissile material and the technical role of making effective use of it, the question arises as to the time they might need to get ready. The period would depend on a number of factors, such as the form and nature of the material acquired and the form in which the terrorists proposed to use it; the most important factor would be the extent of the preparation and practice that the group had carried out before the actual acquisition of the material. To minimize the time interval between acquisition and readiness, the whole team would be required to prepare for a considerable number of weeks (or, more probably, months) prior to acquisition. With respect to uranium, most of the necessary preparation and practice could be worked through using natural uranium as a stand-in.

The time intervals might range from a modest number of hours, on the supposition that enriched uranium oxide powder could be used as is, to a number of days in the event that uranium oxide powder or highly enriched (unirradiated) uranium reactor fuel elements were to be converted to uranium metal. The time could be much longer if the specifications of the device had to be revised after the material was in hand. For plutonium, the time intervals would be longer because of the greatly increased hazards involved (and the absolute need of foreseeing, preparing for, and observing all the necessary precautions). In addition, although uranium could be used as a stand-in for plutonium in practice efforts, there would be no opportunity to try out some of the processes required for handling plutonium until a sufficient supply was available.

To achieve a minimum turnaround time, the terrorists would, before acquisition, have to decide whether to use the material as is or to convert it to metal. They would have to make the decision in part in order to proceed with the design considerations, in part because the amounts needed would be different in the two cases, and in part to obtain and set up any required equipment.

For the first option—using oxides without conversion to metal—the terrorists would need accurate information in advance concerning the physical state, isotopic composition, and chemical constituents of the material to be used. Although they would save time by avoiding the need for chemical processing, one disadvantage (among others) is the requirement for more fissile material than would be needed were metal to be used. This larger amount of fissile (and associated) material would require a larger weight in the assembly mechanism to bring the material into an explosive configuration.

As to the second option—converting the materials to metal—a smaller amount of fissile material could be used. However, more time would be needed and quite specialized equipment and techniques—whether merely to reduce an oxide to the metal or to separate the fissile material from the cladding layers in which it is pressed or sintered in the nuclear fuel elements of a research reactor, for example. The necessary chemical operations, as well as the methods of casting and machining the nuclear materials, can be (and have been) described in a straightforward manner, but their conduct is most unlikely to proceed smoothly unless in the hands of someone with experience in the particular techniques involved, and even then substantial problems could arise.

The time factor enters the picture in a quite different way. In the event of timely detection of a theft of a significant amount of fissile material—whether well suited for use in an explosive device or not—all relevant branches of a country's security forces would immediately mount an intensive response. In addition to all the usual intelligence methods, the most sensitive technical detection equipment available would be at their disposal. As long as thirty-five years ago, airborne radiation detectors proved effective in prospecting for uranium ore. Great improvements in such equipment have been realized since. A terrorist group would therefore have to proceed deliberately and with caution to have a good chance of avoiding any mishap in handling the material, while at the same time proceeding with all possible speed to reduce their chance of detection.

In sum, several conclusions concerning crude devices based on early design principles can be made.

1. Such a device could be constructed by a group not previously engaged in designing or building nuclear weapons, providing a number of requirements were adequately met.

2. Successful execution would require the efforts of a team having knowledge and skills additional to those usually associated with a group engaged in hijacking a transport or conducting a raid on a plant.

3. To achieve rapid turnaround (that is, the device would be ready within a day or so after obtaining the material), careful preparations extending over a considerable period would have to have been carried out, and the materials acquired would have to be in the form prepared for.

4. The amounts of fissile material necessary would tend to be large—certainly several, and possibly ten times, the so-called formula quantities.

5. The weight of the complete device would also be large—not as large as the first atomic weapons ($\sim$10,000 pounds), since these required aero dynamic cases to enable them to be handled as bombs, but probably more than a ton.

6. The conceivable option of using oxide powder (whether of uranium or plutonium) directly, with no postacquisition processing or fabrication, would seem to be the simplest and most rapid way to make a bomb. However, the amount of material required would be considerably greater than if metal were used. Even at full crystal density, the amounts are large enough to appear troublesome: $\sim$55 kg (half bare crit) for 94 percent uranium oxide and $\sim$17.5 kg for plutonium oxide. However, the density of the powder as acquired is nowhere close to crystal density. To approach crystal density would require a large and special press, and the attempt to acquire such apparatus would constitute the sort of public event that might blow the cover of a clandestine operation. Besides, the time required for processing with such a press would preclude a rapid turnaround. Even to achieve densities a little above half of crystal would require some pressing apparatus (not as conspicuous as a large press and conceivably obtainable quietly), but time would again be required to process material quantities of perhaps three or four times those above. The densities available in powder without pressing are not well determined but are quite low, probably in the range of 3 to 4 gm/Cc, although possibly lower.

Within the confines of the crude design category—that of a device guaranteed to work without the need for extensive theoretical or experimental demonstration—an implosion device could be constructed with reactor-grade plutonium or highly enriched uranium in metal or possibly even oxide form. The option of using low-density powder directly in a gun-type assembly should probably be excluded on the basis of the large material requirements.

There remains the possibility of using a rather large amount of oxide powder (tens of kilograms or possibly more) at low density in an implosion-type assembly and simply counting on the applied pressure to increase the density sufficiently to achieve a nuclear explosion. Some sort of workable device could certainly be achieved in that way. However, obtaining a persuasive determination of the actual densities that would be realized in a porous material under shock pressure (and hence of the precise amount of material required) would be a very difficult theoretical (and experimental) problem for a terrorist team. In fact, solving this problem does not belong in the crude design category. Still, a workable device could be built without the need for extensive theoretical or experimental demonstration.

The amount of low-density oxide powder required for a small, crude, implosion-type device is far larger than previously suggested by Theodore Taylor; his view has changed only as to the feasibility of a small, crude device such as terrorists might attempt to build with a single small container of plutonium oxide powder seized from a fuel fabrication plant. We agree, however, that a crude implosion device could be constructed with reactor-grade plutonium or highly enriched uranium in metal or possibly even in oxide form.

7. Devices employing metal in a crude design could certainly be constructed so as to have nominal yields in the 10 kiloton range—witness the devices used in 1945. By nominal yield is meant the yield realized if the neutron chain starts after the assembly is complete and the fissile material is at or near its most supercritical configuration: projectile fully seated in the target for the gun-type device or all the material compressed in the implosion device. In all such systems, there is an interval between the moment when the fissile material first becomes critical (projectile still on its way to its destination, or only a small part of the material compressed) and the time it reaches its intended state. During this interval, the degree of supercriticality is building up toward its final value. If a chain reaction were initiated by

neutrons from some source during this period, the yield realized would be smaller—possibly a great deal smaller—than the nominal yield.

Such an event is referred to as preinitiation (or sometimes predetonation). Obviously, the longer is this interval or the greater is the neutron source in the active material, the larger is the probability of experiencing a preinitiation. The neutron source in even the best plutonium available (lowest Pu-240 content) is so large and the time interval for a gun-type assembly with available projectile velocities (~1000 ft./sec.) is so long that predetonation early in this time interval is essentially guaranteed. For this reason, plutonium cannot be used effectively in a gun-type assembly. The neutron source in enriched uranium is several thousand times smaller than in the plutonium referred to, so that uranium can be used in a gun-type assembly (with available projectile velocities) and have a tolerable preinitiation probability. For this to be true, it is necessary to have rather pure uranium metal, since even small amounts of some chemical impurities can add appreciably to the neutron source. The source in uranium oxide, for example, may be ten or so times larger than in pure metal; the source in reactor-grade plutonium may be ten or more times larger than in weapon-grade plutonium. However, reactor-grade plutonium can be used for making nuclear weapons.

If the assembly velocities (of the projectile or material driven by an implosion) are quite low, the earliest possible preinitiation could lead to an energy release (equivalent weight of high explosive) not many times larger than the weight of the device. If the velocities are quite high (so that the degree of supercriticality increases appreciably during the very short time it takes the neutron chain to build up), the lowest preinitiation yield may still be in the 100-ton range, even in a crude design. Reductions in the weight of the assembly-driving mechanism (whether gun-firing apparatus or amount of high explosive) will, other things being equal, tend to result in lower assembly velocities. The considerations outlined will put some limits on what may be decided to be desirable in connection with a crude design.

8. There are a number of obvious potential hazards in any such operation, among them those arising in the handling of a high explosive; the possibility of inadvertently inducing a critical configuration of the fissile material at some stage in the procedure; and the chemical

toxicity or radiological hazards inherent in the materials used. Failure to foresee all the needs on these points could bring the operation to a close; however, all the problems posed can be dealt with successfully provided appropriate provisions have been made.

9. There are a number of other matters that will require thoughtful planning, as well as care and skill in execution. Among these are the need to initiate the chain reaction at a suitable time and for some reliable means to detonate the high explosive when and as intended.

10. Some problems that have required a great deal of attention in the nuclear-weapons program would not seem important to terrorists. One of these would be the requirement (necessary in connection with a weapons stockpile) that devices have precisely known yields that are highly reproducible. Terrorists would not be in a position to know even the nominal yield of their device with any precision. They would not have to meet the extremely tight specifications and tolerances usual in the weapons business, although quite demanding requirements on these points would still be necessary. Similarly, in connection with a stockpile of weapons, much attention has been given to one-point safety: the assurance that no nuclear yield would be realized in the event of an unplanned detonation of the high explosive, such as might occur in the case of an accident or fire. To ensure the safety of bystanders, this requirement has been deemed important in the context of a large number of devices widely deployed and subject to movement from place to place by a variety of transport modes and by a series of handling teams. Terrorists would not be concerned with this problem, although they would still have a great interest in the safe handling of their device.

11. Throughout the discussion, it has been supposed that the terrorists were home grown. It is conceivable that such an operation could be sponsored by another country, in which case some of the motivation, technical experts, and muscle men might be brought in from outside. This difference would not change the problems that would have to be addressed or the operations required, but it could increase the assurance that important points are not overlooked. It might also provide the basis for considering a sophisticated design rather than a crude type.

More Sophisticated Devices

Most of the schematic drawings that are available relate to the earliest, most straightforward designs and indicate in principle how to achieve a fission explosion, without, however, providing the details of construction. Since 1945, notable reductions in size and weight, as well as increases in yield, have been realized. Schematic drawings of an entirely qualitative sort are also available that indicate the nature of some of the principles involved in these improvements.

Merely on the basis of the fact that sophisticated devices are known to be feasible, it cannot be asserted that by stealing only a small amount of fissile material a terrorist would be able to produce a device with a reliable multi-kiloton yield in such a small size and weight as to be easy to transport and conceal. Such an assertion ignores at least a significant fraction of the problems that weapons laboratories have had to face and resolve over the past forty years. It is relevant to recall that today's impressively tidy weapons came about only at the end of a long series of tests that provided the basis for proceeding further. For some of these steps, full-scale nuclear tests were essential. In retrospect, not every incremental step taken would now seem necessary. Indeed, knowing only that much smaller and lighter weapons are feasible, it is possible at least to imagine going straight from the state of understanding in 1945 to a project to build a greatly improved device. The mere fact of knowing it is possible, even without knowing exactly how, would focus terrorists' attention and efforts.

The fundamental question, however, would still remain: that of whether the object designed and built would or would not actually behave as predicted. Even with their tremendous experience, the weapons laboratories find on occasion that their efforts are flawed. Admittedly, weapons designers are now striving to impose refinements on an already highly refined product, but they have had to digest surprises and disappointments at many points along the way.

For persons new to this business, as it may be supposed a terrorist group is, there is a great deal to learn before they could entertain any confidence that some small, sophisticated device they might build would perform as desired. To build the device would require a long course of study and a long course of hydrodynamic experimentation.

To achieve the size and weight of a modern weapon while maintaining performance and confidence in performance would require one or more full-scale nuclear tests, although considerable progress in that direction could be made on the basis of non-nuclear experiments.

In connection with an effort to reduce overall size and weight as far as possible, it would be necessary to use fissile material in its most effective form, plutonium metal. Moreover, while reducing the weight of the assembly mechanism, which implies reducing the amount of energy available to bring the fissile material into a supercritical configuration, it would not be possible at the same time to reduce the amount of fissile material employed very much. In this case, the amount of fissile material required in the finished pieces would be significantly larger than the formula quantity. Alternatively, in an implosion device without a reduction in weight and size, it would be possible to reduce the amount of nuclear materials required by using more effective implosion designs than that associated with the crude design.

In either case—a small or a large sophisticated device—the design and building would require a base or installation at which experiments could be carried out over many months, results could be assessed, and, as necessary, the effects of corrections or improvements could be observed in follow-on experiments. Similar considerations would apply with respect to the chemical, fabrication, and other aspects of the program.

The production of sophisticated devices therefore should not be considered to be a possible activity for a fly-by-night terrorist group. It is, however, conceivable in the context of a nationally supported program able to provide the necessary resources and facilities and an established working place over the time required. It could be further imagined that under the sponsorship of some malevolent regime, a team schooled and prepared in such a setting could be dispatched anywhere to acquire material and produce a device. In such a case, although the needs of the preparation program might have been met, the terrorists would still have to obtain and set up the equipment needed for the reduction to metal and its subsequent handling and to spend the time necessary to go through those operations.

In summary, the main concern with respect to terrorists should be focused on those in a position to build, and bring with them, their own devices, as well as on those able to steal an operable weapon.

J. Carson Mark was a member of the Nuclear Regulatory Commission's Advisory Committee on Reactor Safeguards and of the Foreign Weapons Evaluation Group of the U.S. Air Force. He was a former division leader of Los Alamos National Laboratories' Theoretical Division and served as a consultant to Los Alamos and a number of governmental agencies.

Theodore Taylor is chairman of the board of Nova, Inc., which specializes in solar energy applications. He is a nuclear physicist who once designed the United States' smallest and largest atomic (fission) bombs. He also designed nuclear research reactors. He has served as deputy director (Scientific) of the Defense Atomic Support Agency and as an independent consultant to the U.S. Atomic Energy Commission. He is coauthor (with Mason Willrich) of Nuclear Theft: Risks and Safeguards and is the subject of John McPhee's *The Curve of Binding Energy*.

Eugene Eyster is a former leader of Los Alamos National Laboratories' WX Division, which is responsible for the explosive components of nuclear weapons. A specialist in chemical explosives, he participated in the Manhattan Project.

William Maraman, a specialist in chemical and metallurgical processing of plutonium and uranium, is director of TRU Engineering Co., which does consulting work on transuranic elements. He was at Los Alamos National Laboratories for thirty-seven years, where he was leader of the Plutonium, Chemistry and Metallurgy Group and of the Material Sciences Division.

Jacob Wechsler is a physicist specializing in nuclear explosives. He was a member of the Manhattan Project and was leader of Los Alamos National Laboratories' WX Division, which is responsible for the explosive components of nuclear weapons.

APPENDIX 2

Civilian Highly Enriched Uranium and the Fissile Material Convention

Codifying the Phase-Out of Bomb-Grade Fuel for Research Reactors

by Alan J. Kuperman

Introduction

Because of the availability of basic nuclear-weapons design information in the open literature and even on the Internet, the main obstacle to fabrication of a nuclear weapon today is the acquisition of sufficient weapon-usable fissile material—plutonium or highly enriched uranium (HEU). The degree to which a fissile material convention can prevent the spread of nuclear weapons, therefore, hinges on its ability to limit the production of, and access to, such materials. A convention that prohibits only the *un-safeguarded* production of weapon-usable fissile materials, but allows unlimited production and use of such materials under safeguards, falls short on two grounds: It permits continued production of weapon-usable material ostensibly for civil purposes that could later be quickly converted by states into weapons; and it permits continued civil commerce in fissile materials, perpetuating the risk of their acquisition by terrorist groups for weapons.

Civil commerce in HEU presents a particular concern because such material has a low background radiation level, making it easier to handle and fabricate into nuclear weapons. Indeed, as Manhattan Project physicist Luis Alvarez wrote in his memoirs:

> With modern weapons-grade uranium, the background neutron rate is so low that terrorists, if they had such material, would have a good chance of setting off a high-yield explosion simply by dropping one half of the material onto the other half. Most people seem unaware that if separate HEU is at hand it's a trivial job to set off a nuclear explosion . . . [E]ven a high school kid could make a bomb in short order.[1]

Moreover, civil HEU has historically been used as a fuel in nuclear research reactors, often located on university campuses that lack the physical security measures employed at many nuclear powerplants and government weapons facilities. The threat posed by continued civil commerce in HEU was underscored dramatically by the disclosure that Iraq, in 1990, diverted bomb-grade uranium fuel from safeguarded research reactors for a crash program to build nuclear-weapon components.

Fortunately, an international cooperative effort, starting in the late 1970s, has made great strides in reducing civil commerce in HEU by converting reactors to non-weapon-usable, low-enriched uranium (LEU) fuels, and by eschewing new HEU-fueled reactors. This effort, known as the Reduced Enrichment for Research and Test Reactors (RERTR) program, headquartered at the U.S. Argonne National Laboratory, has laid the groundwork for the total phase-out of civil commerce in HEU for research reactors.

To preserve the progress of the last two decades and eliminate remaining HEU commerce within the next decade, four further steps are needed. First, development work must continue on the ultra-high-density LEU fuels necessary to convert a few high-power research reactors that still require HEU fuel, and the operators of these reactors must fulfill their pledges to convert as soon as suitable LEU fuel is available. Second, all future research reactors must be designed to use LEU fuel. Unfortunately, Germany has violated this principle with its new FRM-II—the first large research reactor designed to use HEU fuel in the Western world since the RERTR program was created more than two decades ago—and this decision must be rectified. Third, the RERTR principle should also be applied to medical-isotope production—which is the only other significant civilian use of HEU besides reactor fuel—by requiring conversion of all such production to reliance on LEU, rather than HEU, targets. Finally, the proposed fissile material convention should be broadened to codify the phase-out of HEU commerce and the moratorium on any new HEU-fueled reactors.

Background

In the late 1970s the international community belatedly came to the realization that the fuel used in many nuclear research reactors—bomb-grade, highly enriched uranium—could be stolen or diverted for nuclear weapons by nations or terrorists. The RERTR program was established in 1978 to develop substitute fuel of higher-density, low enriched uranium, which is not suitable for weapons. As the substitute fuels were developed, existing reactors would be converted to LEU and new reactors would be designed to use LEU. The RERTR program has proved remarkably successful, facilitating the conversion of dozens of reactors worldwide from bomb-grade to non-weapon-usable fuel and sharply reducing international commerce in HEU.

Outside the United States, some forty-two research reactors with power of at least 1 megawatt were built that originally relied on U.S.–supplied HEU fuel.[2] To date, forty either have converted to LEU, are in the process of converting, have pledged to convert as soon as suitable LEU fuel is available, or have shut down—which has enabled a sharp decline in U.S. HEU exports.[3] Because the United States has historically been the major exporter of fresh HEU for civilian use, and in recent years the sole exporter except for a single Russian transaction with France, this translates into a sharp reduction in total international commerce in bomb-grade uranium. Indeed, as can be seen in figure A1, from 1993 to 1999, there were virtually no exports except minimal quantities for use in production of medical isotopes. HEU exports are currently experiencing a small, temporary resurgence, because the United States has agreed to provide HEU fuel on an interim basis for

Figure A1. RERTR Program Suceeds—U.S. Exports Decline Sharply

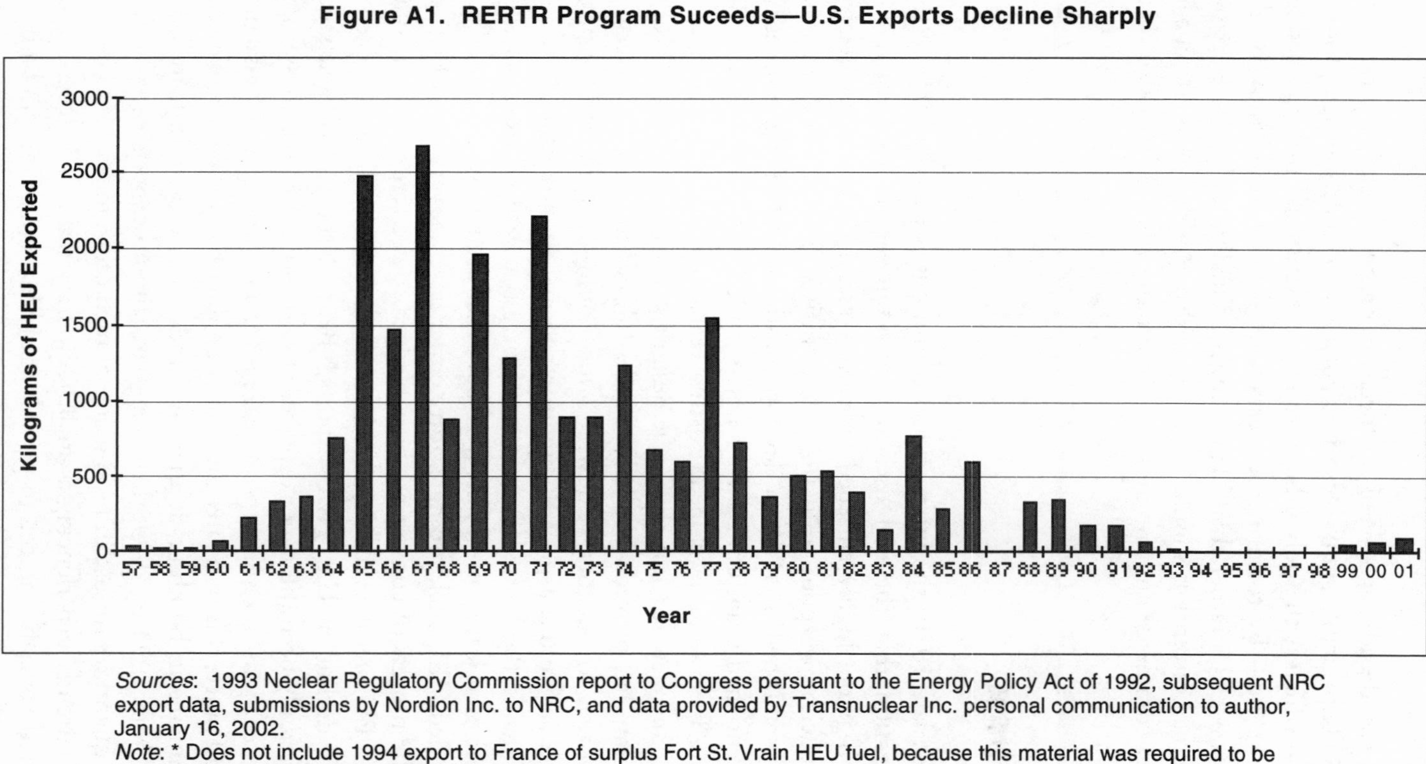

Sources: 1993 Neclear Regulatory Commission report to Congress persuant to the Energy Policy Act of 1992, subsequent NRC export data, submissions by Nordion Inc. to NRC, and data provided by Transnuclear Inc. personal communication to author, January 16, 2002.

Note: * Does not include 1994 export to France of surplus Fort St. Vrain HEU fuel, because this material was required to be blended down LEU as a condition of its export.

a few years to three high-power research reactors in exchange for their pledge to convert as soon as suitable LEU fuel is available.[4]

In addition, the United States has taken steps to reduce its own use of highly enriched uranium. In 1986, the U.S. Nuclear Regulatory Commission ordered the conversion of all licensed, domestic research reactors. Of the twenty-five such reactors operating at the time, ten already have been converted, six are in the process of converting, five ceased operation prior to conversion, three require development of higher-density LEU fuels to enable conversion, and one is a very low power (100kw) private reactor that does not require fresh HEU fuel and is not scheduled for conversion.[5] The U.S. Department of Energy also operates two of its own HEU-fueled reactors, which are not licensed by the NRC. One of these, the HFIR, cannot convert to currently available LEU fuels, according to a recent feasibility study. The other, the ATR, has proposed but not yet carried out its conversion feasibility study. Overall, at least thirty-three reactors located in or supplied by the West already have completed conversion from HEU to LEU.[6]

In recent years, the United States also has entered into agreements with Russia and China to work on conversion of research reactors operating in, and supplied by, these countries. However, progress in achieving actual conversions of these reactors has been painfully slow, which indicates a need for renewed initiative in Washington, Moscow, and Beijing. The United States also is developing a system for production of molybdenum-99 for medical isotopes using LEU targets rather than HEU targets, which currently constitute the main civilian demand for bomb-grade uranium other than reactor fuel.

The key to the RERTR program's success has been two core tenets: spent-fuel return and universality. The U.S. policy of accepting the return to the United States of U.S.–origin spent fuel (both LEU and HEU) through 2009, from operators cooperating with the RERTR program, is based on three grounds. First, it reduces the vulnerability of spent HEU fuel to theft or diversion for nuclear weapons. Second, it honors longstanding U.S. commitments, made originally when the fuel was exported. Third, it helps induce cooperation with the RERTR program, by providing a financial and political incentive for operators to convert their reactors—i.e., they do not have to find an alternate disposal method for their spent fuel until after 2009. The take-back policy extends to LEU spent fuel in order to avoid imposing on reactor operators an additional, perverse penalty for conversion from HEU to LEU—i.e., loss of the right of spent fuel return. The stipulation that spent fuel would be accepted only from reactor operators cooperating with the RERTR program was added—at the recommendation of the Nuclear Control Institute—when the take-back program was renewed in 1996, in order to induce a few remaining foreign operators to convert their reactors. This provision, in combination with the U.S. Energy Policy Act of 1992 that required such cooperation as a condition of interim HEU supply, has helped in obtaining conversion commitments from the operators of the HFR-Petten in the Netherlands, the BR-2 in Belgium, the ILL-Grenoble in France, and the FRJ-2 in Germany.

The principle of universality also has been crucial to the RERTR program by eliminating discrimination as potential grounds for noncooperation. The principle has been applied to the program's three fundamental premises: (1) those reactors that can convert to existing LEU fuel must do so; (2) for remaining reactors, advanced fuel will be developed, to which they must convert when it is successfully qualified; and (3) no new reactors will be constructed to use HEU fuel. Reactor operators have been willing to convert to non-weapon-usable fuel—and to accept the economic and performance penalties of doing so—because the universality principle guaranteed that they would not be put at a competitive disadvantage with respect to neutron research, medical-isotope production, or other reactor activities.

In keeping with this principle, at least thirteen large new research reactors (with power at least 1 MW) constructed since 1980 have commenced operations with LEU fuel, including a 20-megawatt reactor in Japan, 30-megawatt reactors in South Korea and Indonesia, and two U.S. research reactors. (See figure A2.) In addition, another eight research reactors have been designed to use LEU fuel, including the next generation of high-power reactors in China and France.[7] In the same vein, in 1995, the United States abandoned plans for a new HEU-fueled research reactor, the Advanced Neutron Source, despite the pleas of U.S. neutron researchers. The Clinton administration stated at the time that it made this decision at least partly because the bomb-

Figure A2. Large Reactors(>1MW) Constructed after 1980

LEU			
Country	**Construction Start**	**Power (MW)**	**Name**
Peru	1980	3	RP-10
Malaysia	1981	1	Triga
Bangladesh	1981	3	Triga
Indonesia	1983	30	RGS-GAS
Japan	1985	20	JRR-3M
China	1986	1	PPR
China	1986	5	NHR-5
South Korea	1987	30	Hanaro
United States	1987	1	McClellan
United States	1987	1	Triga (U. Texas)
Algeria	1987	1	NUR
Canada**	1990	10	Maple 1
Egypt	1993	22	ETRR-2
Canada**	1998	10	Maple 2
Morocco*	1999	2	Triga
Thailand	Not by 2001	10	MPR-10
France	Not by 2001	100	RJH
China	Not by 2001	60	CARR
Taiwan	Not by 2001	20	TRR-11
Canada	Not by 2001	40	CNF
Australia	Not by 2001	20	"Replacement"

HEU			
Country	**Construction Start**	**Power (MV)**	**Name**
Libya	1980	10	IRT-1
USSR***	1982	10	RBT-10/1
USSR***	1983	10	RBT-10/2
China	1986	5	MJTR
Germany*	1996	20	FRM-11

Sources: IAEA and RERTR program.

Notes:

* Reactor is at least partially constructed but has yet to begin operations.

* Reactor has undergone some low-power testing, but has yet to be licensed for full-power operations.

*** These two Soviet-era reactors in Russia did not actually increase commerce in fresh HEU fuel because they utilize fuel that previously was partially irradiated in the neighboring SM-3 reactor. The dates indicated for construction start of these two reactors are taken from the IAEA, but actual construction may have started prior to 1980, as indicated in figure A3, given that the reactors went critical in 1983 and 1984.

n. b. Press reports also indicate possible plans to construct a 10 MW research reactor in Vietnam and Myanmar, but there is no information on their planned fuel.

grade fuel presented "a non-proliferation policy concern."[8] The only reactors built to use HEU fuel after establishment of the RERTR program were during the cold war in communist China, the Soviet Union, and Libya—at a time when these countries were not observing international nonproliferation norms. The proposed FRM-II would be the first large research reactor built to use HEU fuel since 1980 in a country that claims to observe international nonproliferation norms.

Steps Needed to End HEU Commerce Completely

1. *Advanced Fuel Development and Conversion of Remaining Reactors*—The RERTR program, and an independent French effort, are continuing to develop ultra-high-density LEU fuels necessary to convert the few remaining high-power research reactors that still require HEU fuel. LEU fuel of density 6 g/cc is scheduled for qualification in 2004 and density 8 g/cc in 2006. These fuels should enable conversion of all remaining HEU-fueled reactors around the end of the decade. In addition, there is one HEU-fueled reactor that can convert to existing LEU fuel but so far is not scheduled to do so, the Safari I in South Africa. International leverage is limited, because the operator has its own stocks of HEU (mostly from South Africa's dismantled nuclear weapons), but efforts should be made to bring the operator into conformance with the international norm.

2. *No New HEU-Fueled Reactors*—As noted, Germany's new FRM-II, scheduled to begin operation in 2002, would violate a two-decade international moratorium against new HEU-fueled reactors. As such, it would threaten the RERTR regime in several ways. First, it would generate commerce in more than 400 kgs of bomb-grade HEU fuel in its first decade. German officials recently announced a plan to lower the fuel's enrichment after the first decade—from 93 percent to 50 percent—but even this moderately lower enriched fuel would still qualify as HEU and be usable for weapons. Second, the FRM-II has contracted with Russia to supply the HEU fuel, which encourages Russia to view HEU exports as a lucrative business and to look for additional customers. Increased availability of HEU, combined with Germany's violation of the RERTR program's fundamental principle

of universality, could cause other reactor operators to abandon their planned conversions to LEU or, if they already have converted to LEU, to convert back to HEU. In addition, when other countries build new research reactors, as Vietnam and Myanmar have proposed to do, they may well demand the right to use HEU fuel based on the German precedent. All of this additional HEU commerce would increase risks of nuclear terrorism and nuclear proliferation, just when the world should be reducing such risks in the wake of the terrorist attacks of September 11, 2001.

Fortunately, technical studies by Argonne National Laboratory conclude that the FRM-II can be redesigned to produce equivalent experimental performance using LEU fuel, and would actually enjoy a slightly increased fuel-cycle length.[9] This redesign can be achieved *using LEU fuel that exists today and has been qualified since 1988*—without additional fuel development. Conversion to LEU need not entail substantial delays in completion of the reactor or major increase in cost.[10] The reactor could be redesigned to use already-qualified LEU fuel within a year, and relicensing this new design might require another year. The new design would require a slightly larger reactor core, so that any already completed work constraining the size of the core would have to be modified. Not a single experiment has been identified that would be precluded by conversion to LEU. An LEU-fueled FRM-II also would be no less safe overall than the current HEU design. For all these reasons, the Nuclear Control Institute recently urged German prime minister Gerhard Schroeder to ensure that the FRM-II is converted to LEU prior to its startup.[11]

3. *Convert Medical Isotope Production to LEU*—While the RERTR program has successfully been reducing HEU commerce for reactor fuel, the other major civil application of HEU has been increasing—for use as targets in production of medical radio-isotopes. Such production can and should be converted to LEU. Australia already produces isotopes with LEU, and the RERTR program has enabled Indonesia also to develop successfully an LEU target for medical-isotope production. (In addition, Argentina has worked with the RERTR program toward developing another type of LEU target.) Because of the moderate economic penalties associated with converting medical-isotope production from HEU to LEU targets, the best chance of

Figure A3. International Moratorium on Building HEU-Fueled Reactors since 1980
(Germany Slated to Join Libya and China as Violator)

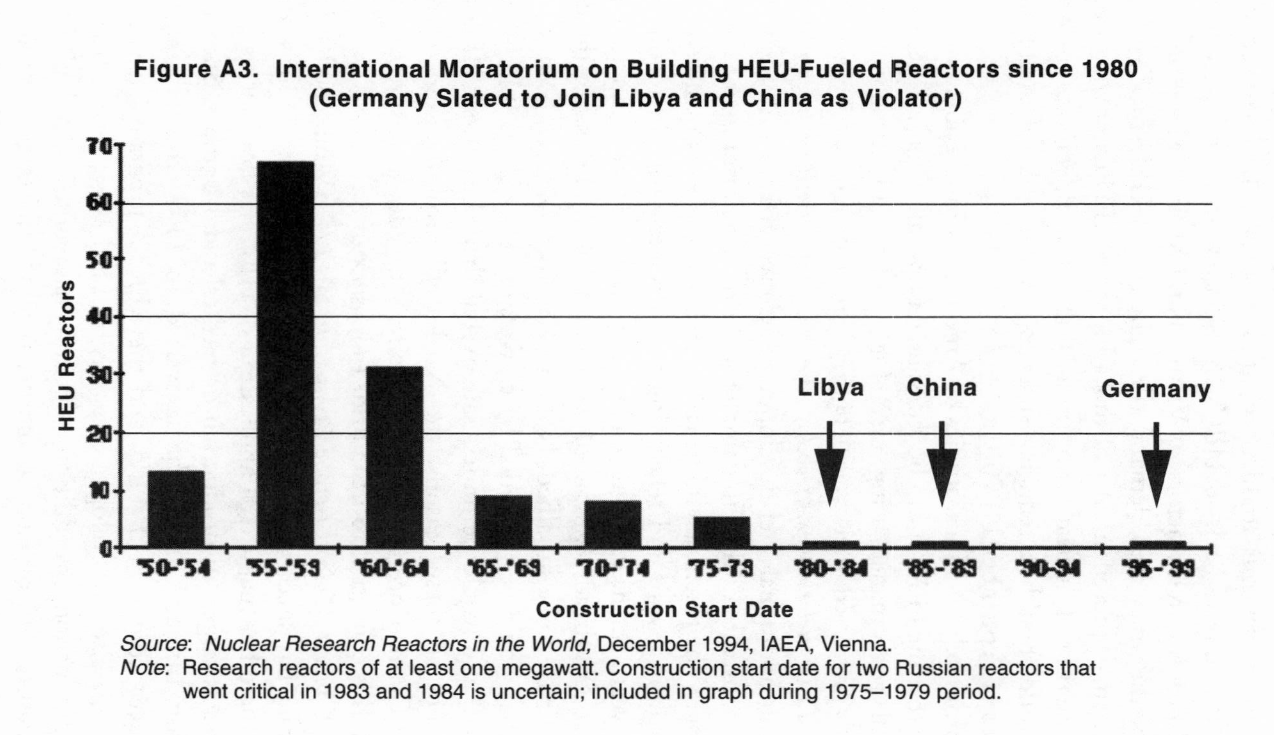

Source: *Nuclear Research Reactors in the World,* December 1994, IAEA, Vienna.
Note: Research reactors of at least one megawatt. Construction start date for two Russian reactors that went critical in 1983 and 1984 is uncertain; included in graph during 1975–1979 period.

Figure A4. Status of Conversion of U.S. Reactors and Foreign Reactors Using U.S. Fuel (Does Not Include Chinese- and Russian-Supplied Reactors)

Location	No. of Reactors	Converted, Converting or Shut Down	Unable to Convert to Existing LEU Fuel	Able to Convert to Existing LEU Fuel but Refusing
United States (at least 1 MW and all university reactors)	22	17	5*	0
Foreign (at least 1 MW) using U.S.–origin HEU	42	38	3**	1***

Notes:

* DOE's ATR reactor (at INEL) and HFIR (at ORNL), the Department of Commerce's NIST (formerly NBSR) reactor, and the university reactors at MIT and University of Missouri–Columbia.

** Belgium's BR-2 reactor and France's ILL-Grenoble and Orphee reactors. The first two have pledged to convert as soon as suitable LEU fuel is available.

*** South Africa's Safari I reactor.

n.b. U.S. university reactors are being converted even if they had low-power (less than1MW) and lifetime HEU cores that did not require fresh fuel. This is in recognition of the extreme vulnerability of university reactors to theft, due to traditionally lax security on most campuses. Other low-power reactors in the United States and elsewhere are not now planned for conversion under the RERTR program, because they do not require fresh shipments of HEU.

getting producers to convert is if they all sign a pledge to convert, so that no producer can gain a competitive advantage by continuing to use HEU. The Nuclear Control Institute has proposed such a pledge and has worked with the RERTR program, the U.S. State Department, and several producers to achieve its universal adoption.[12] The world's largest isotope producer, Canada's Nordion Inc., has indicated its intention to convert to LEU, but so far the other major producers have not.

Conclusion

The RERTR program is one of the unsung heroes of the international nuclear non-proliferation regime and a major bulwark against nuclear terrorism. Since 1978, the program has made great progress in reducing HEU commerce. If the international community provides its full support, the RERTR program can within the decade fulfill its goal of eliminating civil commerce in bomb-grade uranium. However, if Germany operates a new reactor with HEU fuel, South Africa refuses to convert its reactor to available LEU fuel, or medical-isotope producers refuse to convert their processes to LEU targets, the progress of the RERTR program would be seriously undermined and a resurgence of HEU commerce could soon follow. Any unnecessary commerce in bomb-grade uranium represents an unacceptable security risk, especially in light of the increased terrorist threat environment after September 11, 2001. The fissile material convention presents an ideal opportunity to lock in the gains of the RERTR program—by broadening the proposed convention to codify both the phase-out of HEU commerce and the international consensus against construction of new HEU-fueled reactors.

APPENDIX 3

Reactor-Grade Plutonium's Explosive Properties

J. Carson Mark

These comments relate to the question of whether a terrorist organization or a threshold state could make use of plutonium recovered from light-water reactor fuel to construct a nuclear explosive device having a significantly damaging yield. Three aspects of this question will be discussed separately:

I. Criticality Properties of Reactor-grade Plutonium;
II. Effects of Predetonation on Yield Distribution;
III. Some of the Problems Confronting a Terrorist Organization.

Finally, several conclusions are noted in IV.

This chapter is being written at this time because questions appear to persist in some nonproliferation policy circles about whether a bomb really could be made from reactor-grade plutonium of especially high burnup and whether the task is too daunting for a threshold state or terrorist group, even if it is technically feasible.

I. Criticality Properties of Reactor-grade Plutonium

The original implosion assembly system used in the Trinity test in 1945 was capable of obtaining 20 kilotons from weapon-grade plutonium. In the weapons tests conducted in 1948 it was shown that an assembly system of the same size could also handle U-235 effectively. From the discussion below it may be seen that such an assembly system would be capable of bringing reactor-grade plutonium of any degree of burnup to a state in which it could provide yields in the multi-kiloton range. The original implosion system had a diameter of less than 5 feet, including an outer aerodynamic case. Thus, it does not (as was recently suggested) require a "device of the dimensions of a fair sized room" to handle reactor grade plutonium. Moreover, it is well known that the design of the first implosion system was quite conservative, and that there are a number of straightforward improvements that could be implemented to reduce the size of the device on the basis of laboratory-type experiments without having to resort to nuclear tests.

Discussion

In addition to the isotope Pu-239, the plutonium extracted from spent LWR fuel may contain appreciable fractions of other plutonium isotopes formed as a result of successive neutron capture or n-2n reactions. At very low burnup levels the fractional amounts of the secondary isotopes are very small. At a level of a few thousand megawatt-days per metric ton (MWD/MT), for example, the fraction of Pu-240 may be a few percent of the total plutonium, with the fraction of Pu-241 being approximately an order of magnitude smaller, and that of Pu-242 an order of magnitude smaller still. At higher burnups these fractional amounts increase so that at a very high level (≈50,000 MWD/MT or so—about as high as current interest appears to extend) a pattern of the following general sort could be approached: (Pu-239: Pu-240: Pu-241: Pu-242) = (.40: .30: .15: .15:).

Other plutonium isotopes would also be present, but in relatively small amounts. The most prominent of these would be Pu-238, which could reach a level of a few percent in very high burnup material. This would not have a significant effect on critical masses. But because of its relatively short half-lives for alpha decay and spontaneous fission, the amount of Pu-238 might need to be taken into account in determining the alpha activity or neutron source in plutonium from highly exposed reactor fuel.

Each of the plutonium isotopes is sufficiently fissionable that the separated isotope in metal form could provide a bare critical mass, so that a bare critical assembly could be made with plutonium metal no matter what its isotopic composition might be. The odd isotopes (Pu-239 and 241) are both "fissile"—that is, fission may be induced in them by neutrons of any energy, whether slow or fast. Their fission cross sections differ in detail but are similar enough that their bare critical masses are nearly equal, being about 15 kg in δ-phase metal ($\rho = 15.6$ g/cc). The isotope Pu-238 is "fissionable"—that is, only neutrons with energy above some threshold can induce fission. However, the Pu-238 threshold is at some quite low energy, and its fission cross section above about 0.5 MeV is larger than that of Pu-239. In spite of producing fewer neutrons per fission (2.75 vs. 3.0), the bare critical mass of Pu-238 in δ-phase metal is also ≈15 kg.

For Pu-240 the fission threshold is at a few hundred kilovolts; but above 1 MeV the fission cross section, though smaller than that for Pu-239, is larger than that for U-235. The number of neutrons per fission (≈3) is the same as that for Pu-239, 241, and 242. The bare critical mass of Pu-240 in δ-phase metal is about 40 kg. This is smaller than that for 94 percent U-235 in uranium metal at normal density ($\rho = 18.7$ g/cc), which is ≈52 kg. Thus Pu-240 is a significantly more effective fissionable material than 94 percent U-235 in a metal system. It should be noted, however, that this relative superiority would not carry over to the same extent for these materials in the form of oxides. In PuO_2 or U-235 O_2 the average energy of the neutrons is reduced appreciably by their scattering on oxygen. In a Pu-240 O_2 system, therefore, some fraction of the neutrons in the spectrum applicable to a metal system will be moved to energies near or below the Pu-240 threshold where the Pu-240 fission cross section is poor, whereas the fission cross section of U-235 holds up for such lower neutron energies.

At energies above 1 MeV, the fission cross section of the isotope Pu-242 is quite similar to that of Pu-240, but it is a less effective fissionable material because its fission threshold is about a hundred keV higher. The bare critical mass of Pu-242 in δ-phase metal has been calculated to be ≈177 kg. To bring this more in line with the other isotopes, one can think of replacing the Pu-242 component with a new component consisting of a 50–50 mixture of Pu-242 and Pu-241, with the material taken from the Pu-241 fraction since there is enough of that to supply what's necessary even at the extreme high burnup level considered. The fission cross section of this new component, which is the average of the cross sections of Pu-241 and Pu-242, is quite similar to that of Pu-240 in the range of energies above 1 MeV, and considerably larger at lower energies where the Pu-240 and 242 cross sections fall away, while that for Pu-241 does not. Thus, the material of the new component is superior to Pu-240, which in turn is superior to U-235. At all burnup levels, then, the critical mass of reactor-grade plutonium is intermediate between that of Pu-239 and U-235.

By the use of a reflector a few inches thick, the critical mass of all these materials can be reduced by a factor of two, or so, below the bare critical mass; and, at least provided the reflector is of some heavy metal so as not to moderate the neutrons to an important extent, the relative ranking of the critical masses will be preserved.

II. Effects of Predetonation on Yield Distribution

One week after the first fission explosion on July 16, 1945, Robert Oppenheimer wrote to General Leslie Groves' deputy and described the expectations concerning the use of the Trinity device in combat. He said: "The possibility that the first combat plutonium Fat Man will give a less than optimal performance is about 12 per cent. There is about a 6 per cent chance that the energy release will be under 5000 tons, and about a 2 per cent chance that it will be under 1000 tons. It should not be much less than 1000 tons unless there is an actual malfunctioning of some of the components." One week later General Groves wrote to the Chief of Staff: "There is a definite possibility, 12 per cent rising to 20 per cent, as we increase our rate of production at the Hanford Engineer Works, with the type of weapon tested that the blast will be smaller due to detonation in advance of the optimum time. But in any event, the explosion should be on the order of thousands of tons."

Evidently both Oppenheimer and Groves were referring to what will be identified in the following discussion as the "fizzle yield"; that is, the smallest nuclear yield this particular device would provide. They do not state a value for this yield; but in view of their saying "it should not be much less than 1000 tons" it may be presumed that they were thinking of some value like 700 tons or so. The effect of using reactor-grade plutonium in this assembly instead of the high purity plutonium used in 1945 would be to increase the probability that the yield realized would fall short of the levels mentioned by Oppenheimer, but it would not greatly change the actual value of the fizzle yield—which would always be equaled, or exceeded.

In the following discussion some indication is given of the differences between plutonium and highly enriched uranium with respect to pre-detonation and fizzle yields.

Discussion

In any supercritical system, the number of neutrons, the rate of fission, and the level of energy generated increase exponentially—that is, they all vary with time in a way which may be written as $e^{\alpha t}$. The value of the time constant α, which is zero in a system which is just critical and in which the neutron population remains constant, may be as large as one, or a few, times 10^8/sec in a highly supercritical metal system of U-235 or Pu-239. Obviously, the value of α increases with the degree of supercriticality (since a smaller fraction of the neutrons escape without causing a fission), with the density of the fissile material (since, with the atoms closer together, the distance and time for a neutron to cause a fission is reduced), with the average neutron velocity (which is higher in metal than in oxide, for example), and with the factors which favor small critical masses.

Independent of the value of alpha, nothing of much consequence occurs in a supercritical system containing only fissile material in the core until the energy level becomes high enough to vaporize all that material. Only then do pressures build up that can force a disassembly or halt the motion which may be driving the assembly toward a more supercritical condition. At about that point the core begins to expand and its density starts to drop, and the value of alpha (as also the degree of super-criticality, and the rate of increase of the neutron population and energy generation) begin to decrease rapidly toward zero (at which point the system is critical and the neutron population and the energy generation rate are at, or near, their maximum) and on to negative values (where the neutrons rapidly leak away, the energy generation rate falls off, and the reaction is over). Typically, most of the energy from the reaction is developed during this disassembly phase.

As indicated by Robert Serber in the "Los Alamos Primer" of April 1943[1] on the basis of an approximate calculation valid only for a small degree of supercriticality, in any particular system the efficiency of the reaction (the fraction of the fissile material actually consumed) will be proportional to the third power of alpha at the time the motion of disassembly first gets well under way. In a core with a mass of 10 kg or so, this stage will be reached when the value of $\alpha \cdot t$ is somewhere between 40 and 45, where t is measured from the time the chain reaction is initiated. If the system is highly supercritical when the chain starts, so that $\alpha \approx 10^8$/sec, say, then the time for $\alpha \cdot t$ to reach a value ≈ 45 will be extremely short, and there could be rather little change in the degree of supercriticality (or alpha) during this time. However, had the chain started much earlier in the assembly process, when the value of alpha was much smaller, there could have been an appreciable change in alpha during this incubation period—while the 45 generations (as they might be called) were being accumulated. In such a case one would consider the $\int \alpha \cdot dt$ (rather than $\alpha \cdot t$) taken from the time the chain started till $\int \alpha \cdot dt = 45$, and the explosion alpha would be the value applying at the end of that period. Clearly, the smallest possible explosion alpha will be that resulting from a chain which started just as the system reached critical (and alpha reached zero) in the course of its assembly. The yield resulting from this situation will be the smallest possible, and has been referred to as the "fizzle" yield.

Oppenheimer's breakdown of probabilities may be rephrased in the following way: namely that, with the implosion assembly system and the high grade of plutonium being used, the probability was 0.88 that a device would survive long enough without a chain being initiated that it would provide the nominal yield; about 0.94 that it would survive long enough that the yield would be greater than 5 kilotons (one quarter of the nominal); about 0.98 that it would survive long enough to provide a yield in excess of one kiloton. Only in 0.02 of all firings would a chain be initiated so early that the energy release would be between the fizzle yield and one kiloton. Were one to change only the strength of the neutron source (which arises from spontaneous fission and alpha-n reactions) while keeping the mass and reactivity of the fissile material and everything else the same, these probabilities would change. Were the neutron source twice as large, for example, the probability of realizing the nominal yield would be only $(0.88)^2$, *U.S.W.* In particular, for sources 10, 20, 30, and 40 times larger than the one which applied at Trinity these probabilities (and the fraction initiated very close to critical) would be as shown in the following table.

	Yield	*Above 5 kt*	*Above 1*	*Fizzle to 1*
Source:				
Trinity	.88	.94	.98	.02
10 X "	.28	.54	.82	.18
20 X "	.08	.29	.67	.33
30 X "	.02	.16	.55	.45
40 X "	.006	.08	.45	.55

The largest of the sources above is most probably larger than that in the most heavily exposed plutonium considered earlier. It will be seen that as the neutron source is increased from a low level to a very high level the distribution of yields realized changes from one in which the nominal yield is the typical yield and very severe predetonation is rare, to one in which the nominal yield is rare (though never completely excluded) and the typical yields are in a band from one to a few times larger than the fizzle yield.

With the improved data and greatly improved calculation capability that have become available in the meantime, the particular values quoted by Oppenheimer in 1945 would no doubt require some revision. The substitution of a somewhat larger mass of reactor-grade plutonium for the high-grade plutonium employed in the Trinity device would also lead to some changes, both in the nominal yield and the fizzle yield. However, the general pattern pictured above would continue to apply: In the same assembly system some mass of reactor plutonium of any grade would (since this assembly system was capable of making effective use of U-235, which is a less reactive material than reactor-grade plutonium) have a nominal yield of $\approx$10kT or more, and an associated fizzle yield of a few percent of its nominal yield—which is to say, some hundreds of tons. Under heavy predetonation the yields realized would most frequently fall in the range of one to a few times the fizzle yield—never less, but occasionally many times larger. Though almost all of these yields are much smaller than the nominal yield, they would nevertheless constitute quite damaging explosions, and are not reasonably dismissed as "duds" as has sometimes been suggested.

As a final comment concerning fizzle yields, it may be noted that the more rapidly the criticality (or alpha) of the fissile material increases after it first becomes critical, the larger the value of alpha at the moment when $\int \alpha \cdot dt = 45$. If, for example, we assume that alpha increases linearly with time, so that $\alpha = k \cdot t$, then, when $\int \alpha \cdot dt = \sqrt{45}$, we have $t = \int(90/k)$ and $\alpha = \int(90 \cdot k)$—which is larger, the larger k may be. Since the efficiency of the fizzle explosion varies as the cube of this value of alpha, the faster the assembly proceeds, the larger the fizzle yield of a given mass of fissile material. From the fact that the Trinity assembly was a very conservative design, it would seem likely that straightforward ways could be found to realize a faster-moving implosion, which could have the effect of increasing fizzle yields to higher levels than those applying above. On the other hand, since the time interval from first critical to complete assembly might be something like 50 times longer in a gun-assembly system than in an implosion—so that the slope of the alpha-curve (the value of k, above) would be much smaller—not only would initiation be essentially guaranteed early in the assembly process even by the neutron source in very high-grade plutonium, but the value of alpha at the earliest possible explosion time would be smaller by a factor of something like $\approx$50, and the fizzle yield would be reduced by a large factor. Thus, not even the best weapons-grade plutonium is of any interest in connection with a gun-type assembly system.

These considerations come out quite differently in connection with highly enriched uranium because the neutron source from spontaneous fission in such material is smaller than that in even the best grades of plutonium by a factor of more than a thousand. In the relatively slow-moving gun-type device one might wish to assemble a couple of critical masses or so, which would imply bringing together something like 50 kg of 94 percent U-235, since the critical mass with a reflector can be about half the bare critical mass of 52 kg. The fizzle yield of such a system would, again, be some uninteresting low value; but, with the very low neutron source that could be realized in this material, the probability of initiating a chain at a very early stage of the assembly process may be small enough to ignore. Indeed, Luis Alvarez, a scientist with the Manhattan Project during its war years, has said, "With modern weapons-grade uranium the background neutron rate is so low that terrorists, if they had such material, would have a good chance of setting off a high-yield explosion simply by dropping one half of the material on to the other half."[3] What he meant by "high-yield" or "good chance" are not explained; but his mere statement calls attention to the fact that highly enriched uranium is in a class by itself.

III. Some of the Problems Confronting a Terrorist Organization

Technical Personnel

Competence and thorough understanding will be required in a wide range of technical specialties. These include: shock hydrodynamics, critical assemblies, chemistry, metallurgy, machining, electrical circuits, explosives, health physics, and others. At least several people who can work as a team will be needed. These will have to be carefully selected to ensure that all necessary skills are covered, but they need not have been previously engaged in designing or building nuclear weapons.

Costs

In addition to support for the personnel over a period adequate for planning, preparation, and execution, a considerable variety of specialized equipment and instrumentation will be required, all or most of which can be obtained through commercial sources.

Hazards

Radiation, criticality, the handling of noxious materials, and explosives all present potential hazards that will have to be foreseen and provided against.

Detection

Assuming the operation is contrary to the wishes of the local national authorities, the organization must exercise all necessary precautions to avoid detection of their activities. They would no doubt be faced by a massive search operation employing the most sensitive detection equipment available should it become known that someone had acquired a supply of material suitable for use as an explosive.

Acquisition

Very early in its planning and equipment procurement phase the organization will need information concerning the physical form and chemical state of the fissile material it will have to work with. This will be necessary before they can decide just what equipment they will need. The isotopic content of the material could be determined by straightforward means. The actual acquisition of the material would probably be the responsibility of a separate task force for which the problems and hazards would be those set by the safeguards and security authorities.

IV. Conclusions

1. Taking "weapon" to signify an object suitable for stockpile by a military organization, then heavily irradiated reactor plutonium would not be attractive for an arsenal of pure fission devices. For that purpose one would wish to have a set of warheads with a reliable known yield.

One would also wish to have objects that could be turned out in a production-line fashion. However, for a terrorist organization acting alone or on behalf of a rogue state, with interests focused on the possible use of one, or a very few, devices, these considerations might be weighed quite differently. In addition, radiation exposures associated with fabrication that might be unacceptable for a sustained activity might not be troublesome for a one-shot operation.

2. It has been suggested that the fact that the United States appears to have made only one experiment using reactor-grade plutonium and has not chosen to adopt it for regular weapon production indicates that such material is of little worth. That is not the correct interpretation. There is, of course, no question but that weapon-grade material is preferable from a design standpoint; and if, as for the United States, one has the option and is paying for the plutonium anyway, one chooses the most advantageous. So would the terrorist if he had a choice. But if he can't get weapon-grade material, he would take whatever he can get, should any be open to him.

3. The technical problems confronting a terrorist organization considering the use of reactor-grade plutonium are not different in kind from those involved in using weapon-grade plutonium, but only in degree. For example, it is of great importance to avoid the inhalation of plutonium dust or vapor; but the provisions which would be adequate for weapon-grade material would require little, if any, modification to be acceptable for reactor-grade material. The hazards and difficulties associated with assembling a device would be less if highly enriched uranium were used.

4. The method of coping with the problems and difficulties of making an explosive device with reactor-grade plutonium is entirely in the hands of the terrorist organization. The information necessary to meet the needs is available, and can be assembled by a properly chosen team of specialists. It cannot be said whether or not they would conclude that the effort involved is within their reach, or "worthwhile," since that depends on many factors known only to them. It can be said that the only point on which established authorities can influence their decision is on that of the acquisition of material. Whether that should be more or less difficult, and whether or not the fact of their successful

acquisition would be known rapidly and with assurance, could be important in this respect.

5. Assuming they do not also have access to a supply of highly enriched uranium, and assuming that the working group in question has been specifically formed to produce a first device in as short a time as possible with a high degree of confidence in obtaining a significant nuclear yield, the amount of material they would have to acquire could scarcely be as small as 5 kg, though it might not have to be very much larger than 10 kg. Even for a working group with the time and the means of conducting an extended series of non-nuclear assembly experiments—circumstances more likely to apply to a group engaged in a national effort by some Nth country than to a terrorist group—an amount of at least several kilograms would be necessary.

6. It has been suggested that rather than trying for an explosive device a terrorist organization might merely set out to disperse a quantity of reactor-grade plutonium in some highly populated location. This would bypass many of the difficult technical problems involved in producing an explosive device; and in this case reactor-grade plutonium, being several times more noxious than weapon-grade, could be the material of choice. However, it is not fully clear what objective would be realized by actually going through with such an action that could not be met as well, or better, by a well-publicized and credible threat. Here, again, the main line of defense available to the authorities is to ensure that the acquisition of such material is difficult, and that they have the means of assuring themselves rapidly whether or not the material claimed to be available is missing.

7. Finally, if methods of separating plutonium isotopes using laser technology (already receiving serious consideration in the United States) should, in the future, come within the reach of many industrial states, then stocks of reactor-grade plutonium would present a much more direct access to proliferation of nuclear weapons than they may appear to do at present.

APPENDIX 4

Are IAEA Safeguards on Plutonium Bulk-Handling Facilities Effective?

Marvin M. Miller

The purpose of this chapter is to assess the effectiveness of international (IAEA) safeguards at peaceful nuclear fuel cycle facilities which handle plutonium in bulk form. There are two facilities of this type: reprocessing plants which extract the plutonium from nuclear fuel irradiated in nuclear reactors, and fabrication plants which process the extracted plutonium into fresh fuel assemblies.

I. Introduction

The rationale for the use of plutonium in the nuclear fuel cycle is the contention that such use is the only means of ensuring the viability of nuclear power as a long-term energy source. That is, the present generation of nuclear power reactors are too inefficient in the use of uranium to sustain a large contribution of nuclear power, say, 1000 GWe worldwide, beyond the next century. The only way to extend this time horizon significantly is to use plutonium-fueled fast breeder reactors. For this reason research, development, and demonstration of breeder reactors and the associated fuel cycle facilities were an early feature of the nuclear power programs in the major industrialized countries.

The vision of nuclear power as an inexpensive and environmentally-benign source of energy has dimmed considerably in recent years mainly because of widespread concerns about reactor safety, as well as the disposal of radioactive wastes, and the possible misuse of peaceful nuclear facilities and materials by both states and sub-national groups for the production of nuclear weapons. In response to this reality and the realization that low-cost uranium resources are abundant, the emphasis of nuclear development and demonstration, particularly in the United States, has shifted from rapid movement to a plutonium breeder economy toward the validation of new reactor concepts which have a higher degree of passive safety than existing reactors.

Despite this, the nuclear establishment in several countries, notably France and Japan, still insists that in the long run the plutonium breeder will be needed. To this end, it is necessary to gain experience in plutonium operations via operation of reprocessing and fuel fabrication plants and use of plutonium fuel in both prototype breeder reactors and existing light water reactors.

The counterargument is that given especially the substantial potential for both increased energy efficiency and wider use of various renewable energy sources, the plutonium breeder will not be needed until well into the twenty-first century (if ever), even if the threat of greenhouse warming limits the use of fossil fuels. From this perspective, the uncertain economic benefits and waste disposal advantages of plutonium reprocessing and recycle in light water reactors are outweighed by the substantial risks of diversion of this material, as well as its potential release to the environment in normal fuel cycle operations, and under accident conditions.

Although the issue of the need for nuclear power in general and plutonium breeders in particular requires the consideration of its comparative economics and environmental impacts as well as its proliferation and terrorism risks, we consider only this last aspect here, and further restrict ourselves to the question of the effectiveness of safeguards at plutonium bulk handling facilities. We begin in the next two sections with a brief sketch of the relevant safeguards background.[1]

II. Safeguards Goals

In contrast to IAEA document INFCIRC 66/REV. 2, which delineates the safeguards system for nuclear facilities in non-NPT states, the corresponding document that was developed to detail the safeguards obligations of states party to the NPT, INFCIRC/153, provides a technical definition of the safeguards objective, namely "the timely detection of the diversion of significant quantities of nuclear materials from peaceful activities . . . and deterrence of such diversion by the risk of early detection." The key terms of this objective were not defined in INFCIRC/153; this task was given to the Standing Advisory Group on Safeguards Implementation (SAGSI) of the IAEA, an advisory group of technical safeguards experts.

SAGSI considered the problem of quantifying the safeguards objective for several years. It identified four terms appearing either explicitly or implicitly in the statement of the objective just quoted as in need of quantitative expression. These were: *significant quantities, timely detection, risk of detection, and the probability of raising a false alarm.* It defined the associated numerical parameters (significant quantity, detection time, detection probability, and false alarm probability) as *detection goals.*

In 1977, SAGSI submitted numerical estimates for these goals to the director of safeguards of the IAEA. A significant quantity (SQ) was defined as "the approximate quantity of nuclear material in respect of which, taking into account any conversion process involved, the possibility of manufacturing a nuclear explosive device cannot be excluded." For plutonium the significant quantity was taken to be 8 kg; for highly enriched uranium (HEU), 25 kg of contained U-235; for low-enriched uranium (LEU), 75 kg of contained U-235.

Detection time (the maximum time that should elapse between a diversion and its detection) should be of the same order of magnitude as conversion time, defined as the time required to convert different forms of nuclear material to the components of a nuclear explosive device. For metallic Pu and HEU, conversion time was estimated as seven to ten days; for pure unirradiated compounds of these materials such as oxides or nitrates, or for mixtures, one to three weeks; for Pu or HEU in irradiated fuel, one to three months; and for low-enriched uranium, one year.

On the basis of common statistical practice, SAGSI recommended a detection probability of 90–95 percent, and a false-alarm probability of less than 5 percent.

The values recommended by SAGSI for the detection goals were carefully described as *provisional guidelines* for inspection planning and for the evaluation of safeguards implementation, not as requirements, and were so accepted by the Agency. However, the view of a sector of the non-proliferation community, which was particularly influential in the United States during the Carter administration, was that unless these goals could be met in practice, safeguards were not effective, and the associated activity, e.g., reprocessing of spent reactor fuel to extract plutonium, posed too great a proliferation risk. This perspective was embodied in the major piece of nonproliferation legislation enacted into law in the United States during the Carter administration, the Nuclear Non-Proliferation Act (NNPA) of 1978. This is particularly true with regard to the importance given to the concept of "timely warning" in the NNPA. "Timely" is taken to be detection of a diversion quickly enough to take diplomatic action to prevent the fabrication and insertion of the diverted material into a first bomb that is otherwise complete. Thus, detection time must be even shorter than conversion time, in order to allow for evaluation and response, e.g., even shorter than one to three weeks for Pu or HEU compounds in unirradiated form.

In the author's opinion, the view that the detection goals should be operational criteria for safeguards effectiveness has a certain logic. In particular, a diversion of less than a significant quantity would not provide enough material for a nuclear explosive, and with regard to timely warning, it would obviously be advantageous to know about a diversion in time to do something about it, that is, before the diverter could assemble a weapon from the diverted material. However, in judging safeguards effectiveness, it is also logical to be able to distinguish between a situation in which the performance of the safeguards system is only marginally worse than the detection goals, and one in which the differences are considerable. A relevant example would be two reprocessing plants in which the minimum amount of diverted plutonium which could be detected with high confidence is 10 kg/yr and 250 kg/yr, respectively. This raises the issue of what other factors besides numerical goals might be taken into account in judging safeguards effectiveness. Before taking up this issue, we first assess the current situation with regard to the capability of the safeguards system in meeting the detection goals, and the prospects for future improvements.

III. Material Accountancy

The paragraph in INFCIRC/153 immediately following the one in which the safeguards objective is defined delineates the methods to be used in timely detection of diversion:

> To this end the Agreement [between the Agency and the State] should provide for the use of materials accountancy as a safeguards measure of fundamental importance, with containment and surveillance as important complementary measures.

Application of material accountancy by the IAEA to the detection of diversion of nuclear material is analogous to a bank examiner's financial audit.[2] First the nuclear facility operator [bank management] must prepare a material balance [financial statement] covering a specified period, e.g., one year, showing that all nuclear material [money] can be accounted for. More specifically, adding the material inputs (I) [credits] and subtracting the removals (R) [debits] from the beginning inventory (BI) [opening balance] gives the amount that should be in the ending inventory (EI) [final balance]. The IAEA inspector [bank examiner] performs an independent check on at least some of the data presented by the facility operator [bank management] to confirm the absence of deliberate falsification.

This procedure works well in the context of the financial audit and at nuclear facilities where the nuclear material is present only in the form of identifiable and countable items, e.g., fuel assemblies at power reactors. That is, if the "books do not balance," it is a clear indication that there has been an unrecorded removal of nuclear material [money] from the facility [bank]. In the parlance of nuclear materials accountancy, a positive value of "terials unaccounted for" or MUF, defined by

$$\text{MUF} = (\text{BI} + \text{I} - \text{R} - \text{EI}), \quad (1)$$

where R includes both product and any lost material indicates a diversion, whereas if MUF = 0 and the operator's data have been authenticated by the inspector, then it is possible to unambiguously conclude that no diversion has occurred.

When eq. (1) is applied to materials in bulk form, a problem arises because it is no longer possible to know any of the terms in the equation exactly. Unlike fuel assemblies, the quantity of bulk materials, such as plutonium in reprocessing and fuel fabrication plants and uranium in fabrication and enrichment plants, can only be measured approximately. As a result, even in the absence of diversion, non-zero values of MUF will be measured, and materials accountancy must rely on statistical tests to distinguish positive values of MUF due to diversion from those due to a chance combination of measurement errors. Whether this is a significant problem in terms of meeting the IAEA's detection goal of detecting an SQ depends on the magnitude of the associated errors—technically specified by the *variance* of MUF, s(MUF)—compared with an SQ.[3] Intuitively, if the "noise" of the measurement process, as specified by s(MUF), is small compared to an SQ, then diversions of material on the order of an SQ should be detectable with both high confidence and only a small probability of a false alarm. Conversely, if s(MUF) greatly exceeds an SQ, then the minimum diversion which can be detected with high confidence and a small false alarm probability will also be much greater than an SQ.

Unfortunately, even if s(MUF) is small as a *percentage* of the quantity of material measured, e.g., = ±1%, in a plant processing large quantities of material, the *absolute value* of s(MUF) will, over a sufficiently long period of time, exceed an SQ.

A relevant example is the planned 800 tonne/yr Rokkasho reprocessing facility at Aomori in Japan. Assuming that: (1) the plant processes spent fuel with an average total plutonium content of 0.9%; (2) the error in measuring the MUF, specified by s(MUF), is dominated by the error in measuring the plutonium input, and is equal to ±1% of this input, and (3) the material balance calculation is done once a year, then the absolute value of s(MUF) = 72 kg of Pu/yr. It is straightforward to show that the minimum amount of diverted plutonium which could be distinguished from this measurement "noise" with detection and false alarm probabilities of 95 percent and 5 percent, respectively, is 3.3 &s(MUF), or 246 kg in this example, equivalent to more than thirty significant quantities.

Besides the fact that the minimum detectable diversion in such a plant greatly exceeds an SQ, the detection time will also exceed the timeliness goals for the various forms of plutonium in the plant, e.g., the one to three week goal for unirradiated plutonium compounds such as the plutonium nitrate product of the plant. There are three reasons for this.

In the first place, while the material balance is measured on a yearly basis, the diversion of 8 kg or more of plutonium could take place at any time following plant startup. (Note that an 800 tonne/yr reprocessing plant might process 4 tonnes of fuel containing 36 kg of plutonium *per day* for 200 days, with the remainder of the time reserved for scheduled and unscheduled maintenance.) Secondly, the determination of the concentration of plutonium in the input and output accountability tanks, as well as in process tanks, currently requires that tank samples taken by the plant operator and given to the IAEA inspector be shipped back to the IAEA analytical laboratory outside Vienna for measurement. Because of stringent national regulations on the shipment of plutonium, this is often a time-consuming process: delays in measuring samples on the order of months are not unusual.

Finally, in the Agency's view, a false accusation of diversion would be extremely serious, and could discredit the safeguards system. Thus, detecting a diversion means, first, detecting a suspicious event, technically an "anomaly," indicative of a *possible* diversion, such as a large MUF or a film picture indicating unreported movement of nuclear material. The Agency then attempts to systematically eliminate all other possible explanations, such as larger than estimated measurement errors, unreported material losses, defective safeguards equipment, etc. This process is apt to be very time-consuming, especially if remeasurement is required, and the greater the degree of certainty that is required, the longer the process will take. Thus, detection in the spirit of the timely warning philosophy cannot in practice be realized both because of the "untimely" nature of the measurement process and also because of the Agency's philosophy of being extremely careful to avoid an unjustified accusation of diversion.

One obvious technical fix is to perform material balance measurements more frequently, e.g., weekly. Assuming that the percentage error remains the same, a shorter measurement period implies a smaller absolute value of (MUF) since the plant throughput is smaller. This increases the potential for both greater detection sensitivity and timeliness in the detection of an abrupt diversion. Unfortunately, making inventory measurements in large plants, particularly reprocessing plants, is time consuming and expensive because it involves a shutdown of the plant and a washout of the process equipment. For this reason, only one or two inventory takings per year would be acceptable to the plant operator.

Recognizing these difficulties, the Agency has tried to remedy them in two ways: (1) by considering the implementation of advanced technical approaches, in particular, near-real-time accountancy (NRTA), wherein material balance measurements are made frequently without shutting down the plant, and greater reliance on containment/surveillance (C/S) measures; and (2) by setting up two other levels of safeguards goals, the *inspection goals,* which are supposed to reflect actual conditions at the facility, requirements prescribed by the safeguards agreements, and capabilities of safeguards measures, and the *accountancy verification goal,* defined as the minimum quantity of diverted material which could be detected with 95 percent and 5 percent detection and false alarm probabilities, respectively, based on achievable measurement uncertainties; currently, the latter is specified by the Agency as a (MUF) of $\pm 1\%$ of the input, e.g., 246 kg of plutonium per year at an 800 tonne/yr plant.

The potential of NRTA and greater reliance on C/S measures for improving the effectiveness of safeguards at bulk-handling facilities is discussed in the next section. As to the remedy of defining alternative safeguards goals, this is widely—and, in the author's view, correctly—perceived as a retreat from the original detection goals, indicative of both the Agency's inability to meet these goals and its unwillingness to admit this fact. The result has been a loss of confidence in IAEA safeguards, particularly in the United States.

In sum, the IAEA's safeguards detection goals cannot be met at large reprocessing and plutonium fuel fabrication facilities using conventional materials accountancy. Although we have followed the common practice of focusing on safeguards in the chemical process area of a reprocessing plant, it is also important to note that there are significant errors in the current measurements of other plutonium-bearing streams in such a plant. In particular, the fuel hulls and the undissolved plutonium which is filtered out of the process stream before it reaches the separation stages both contain on the order of 0.5 percent of the plutonium input, respectively, or 40 kg/yr in an 800 tonne/yr facility. The accuracy with which these streams are measured currently is on the order of 50 percent of their magnitudes, or about 20 kg of plutonium. Thus even if the potential of NRTA is realized in the process area, the goal of detecting an SQ will not be satisfied unless significant improvements in these measurements are also realized. We return to this point in the next section.

IV. The Technical Potential for Improved Safeguards Performance

In this section, we briefly discuss the technical potential for higher detection sensitivity and greater detection timeliness in large plutonium bulk-handling facilities, particularly reprocessing plants. We begin with a discussion of Near-Real-Time Accountancy (NRTA), that is, making measurements of the material balance more frequently than in conventional materials accounting, e.g., weekly instead of yearly.

What makes NRTA practical is the feasibility of making frequent measurements of the plant's plutonium inventory without shutting it down. This is accomplished by actual measurement of plutonium in most of the process equipment, and reliance on estimates of the plutonium content of those vessels which are inaccessible to measurement.

As previously noted, use of NRTA implies a smaller absolute value of s(MUF) since the plant throughput during the material balance period is proportionately smaller. Thus, for the 800 tonne/yr = 4 tonne/day reprocessing plant we have been using as an example, the weekly throughput of plutonium would be 252 kg. At a measurement error s(MUF) = $\pm 1\%$, the minimum diversion which could be detected with detection and false alarm probabilities of 95 percent and 5 percent, respectively, is then about 8 kg.

Weekly material balance takings would also increase the timeliness of diversion detection *if* the measurement samples could be analyzed more quickly. That is, instead of shipping the samples to the IAEA's analytical laboratory outside Vienna, the measurement techniques employed should be amenable to rapid analysis by the IAEA inspectors on-site. Several techniques of this type are in the development and demonstration stage, and their prospects are promising.

The efficiency of NRTA in detecting a *protracted* diversion of plutonium, i.e., diversion of a small amount of plutonium per week, whose cumulative total over many weeks exceeds 8 kg, is not as clear. (Note that conventional materials accountancy does not distinguish between abrupt and protracted diversion since measurements are only made yearly). The argument of NRTA proponents can be summarized as follows. Assume that one has a significant data bank of MUF values for a period during which there was no diversion. Then the deviation of MUF values from zero during this period must be due to measurement error.[4] Thus, in testing another sequence of MUF values for protracted diversion, one can subtract an estimate of the measurement error derived from the diversion-free data from the sequence of MUF values under test. In this manner, one should be able to effectively decrease the magnitude of (MUF), and hence increase the diversion detection sensitivity.

This argument is intuitively appealing, but not entirely convincing, at least not to the author. While there is a considerable literature which seeks to demonstrate the efficacy of various sequential testing procedures in detecting protracted diversion, their basic assumption is the existence of diversion-free MUF data which can be used as a calibration standard. If diversion begins when the plant starts operation and continues as long as safeguards are applied, then there is no such data, and Avenhaus and Jaech have shown that there is no gain in detection sensitivity using NRTA compared with conventional materials accountancy.[5] Moreover, the sequential tests do not give the operator-diverter credit for diversion strategies which are more sophisticated than simply removing a fixed amount of plutonium during successive material balance periods. For example, the operator could also put material into the system during some of the material balance periods. This makes diversion look more like measurement noise, and hence makes it more difficult to detect. It may turn out that a selected menu of different sequential tests may be able to detect all credible protracted diversion scenarios with a detection sensitivity significantly greater than that available from conventional materials accountancy, but this remains to be demonstrated.

Finally, it must be recognized that implementation of NRTA would be labor intensive for both the plant operator and the IAEA, and would provide the latter with a degree of insight into plant operations which, while beneficial from the viewpoint of safeguards effectiveness, might also conflict with the operator's desire to protect proprietary information. Thus, higher safeguards costs as well as some degree of opposition from plant operators are likely.

Regarding the prospects for achieving greater safeguards effectiveness by increased reliance on containment and surveillance (C/S) measures, it should be noted that the IAEA safeguards staff and the safeguards support programs in the member states, particularly the United States, have invested much time and effort over the past fifteen years to develop and implement reliable and effective C/S devices, particularly seals and cameras. There is no doubt that such devices have the potential for fulfilling the role of C/S as an important complement to materials accountancy envisaged for it in INFCIRC/153. For example, the use of cameras to provide surveillance of both the spent fuel pool at a reprocessing plant and the transfer of such fuel to the chop-leach cell can detect attempts to process undisclosed spent fuel in the plant. Another relevant example is the use of seals on the tanks containing the plutonium nitrate product of such a plant to detect unauthorized withdrawals of material.

The problem has always been that, while it should be possible to take credit for C/S measures in judging safeguards effectiveness, no one has figured out a logical way of quantifying this benefit and combining it with the assurance of non-diversion provided by materials accountancy into a combined measure of safeguards effectiveness. Nevertheless, the benefit is real and justifies continued effort to improve the reliability and effectiveness of C/S devices.

The principal caveat to the above is that C/S measures cannot substitute for NRTA, particularly in the process area of a reprocessing plant where it is impractical to monitor the myriad of pipes, valves, pumps, and tanks using such devices.

Finally, more accurate measurements of the plutonium content of waste streams, particularly the hulls and sludges in reprocessing plants, are required to approach the 8 kg/yr detection goal. Various non-destructive assay (NDA) techniques have been used to measure such streams, but the associated errors are large. Active neutron interrogation shows the most promise, but it is also the most difficult to implement.

V. Conclusion

Assuming, optimistically, that: (1) a better measurement of the plutonium in the input accountability tank of a reprocessing plant makes it practical to achieve a s(MUF) = $\pm 0.5\%$ overall in the chemical process area; (2) the use of NRTA on a weekly basis makes it possible to improve on the detection sensitivity of conventional materials accountancy for protracted diversion by a factor of 4, and (3) the plutonium in hulls and sludges can be measured to $\pm 10\%$, then the minimum detectable abrupt and protracted diversions would be about 5 kg and 35 kg, respectively, in an 800 tonne/yr reprocessing plant.

The situation could be somewhat better in a large mixed-oxide (MOX) fuel fabrication plant because the plutonium streams are easier to measure. For example, the plutonium throughput of a 100 tonne/yr MOX facility which fabricates fuel with an average plutonium concentration of 3 percent would be 3000 kg. Assuming a s(MUF) = $\pm 0.3\%$ overall, and monthly NRTA with an "effectiveness gain" of a factor of two for protracted diversion compared to conventional materials accountancy, the minimum detectable abrupt and protracted diversion would be about 3 kg and 15 kg, respectively.

Is the considerable effort that will be required to achieve these results worth it? The proposition that it is not is usually based on the argument that states which voluntarily accept safeguards by joining the NPT are not likely candidates for diversion. Thus, even if the safeguards goals cannot be rigorously achieved, or even reasonably approximated, the proliferation risks are minimal.

The usual counterarguments are that: (1) such political "credit" cannot be given to non-NPT states who do not accept safeguards voluntarily, but only as a condition for technology transfer; (2) not all NPT states have impeccable non-proliferation credentials, and in practice, it is difficult to implement a safeguards system based on subjective, and changing, judgments about proliferation risk. In the author's view, these counterarguments have merit. *However, the most telling argument for more effective international safeguards is its likely impact on domestic safeguards and the risk of sub-national diversion of plutonium.* That is, international safeguards are based on national systems of materials accountancy and control. Without the incentive for greater safeguards effectiveness provided by stringent IAEA safeguards criteria, there might be a tendency to relax domestic safeguards. This, in turn, would increase the risk of subnational diversion, particularly with the collusion of an insider, who was familiar with the weak points of the safeguards system. A historical case in point is that of the plutonium weapons facilities in the United States. The "mind-set" of the U.S. Atomic Energy Commission, under whose auspices these plants were built and operated, was that the first priority was maximum production, and that the insider threat was not credible. As a result, materials accountancy and control left much to be desired, and considerable material was unaccounted for. While all of this might have plated out in the process equipment or been released to the environment rather than being diverted, operation of plutonium-handling facilities in a manner which disregards both the insider threat and environmental hazards is unacceptable.

In sum, technical measures, especially NRTA, but also, more reliable and effective C/S, greater at-plant IAEA measurement capability, and more accurate measurements of the plutonium in waste streams, could lead to a significant improvement in the effectiveness of international safeguards at large plutonium-handling facilities. Implementation of such measures would increase public confidence in the ability of the IAEA to minimize the risks of the use of plutonium in nuclear fuel cycles. Until these measures can be implemented *and* demonstrated, it would be prudent to limit plutonium use to research, development, and demonstration projects. For even if these improvements can be practically achieved, there are still diversion risks as well as environmental hazards associated with large-scale transport of plutonium between reprocessing/fabrication plants and reactors. With the current glut of low-cost uranium, the world can afford to take the time to investigate the feasibility of reactors, including breeders, which are not only safer and make waste disposal more tractable, but also have a higher degree of proliferation and terrorism "sistance" than the standard breeder and its associated fuel cycle. If nuclear power is to have a future, it should be in this direction.

NOTES

Chapter 1: Introduction: Nuclear Power without Proliferation?

1. British Nuclear Industrial Forum, "The BNIF Foresight Programme: An Associate Programme of the National Foresight 2000 Programme," London, November 9, 2000.

2. David H. Albright, "World Inventories of Civilian Plutonium and the Spread of Nuclear Weapons," Nuclear Control Institute Special Report, Nuclear Control Institute, Washington, D.C., 1983.

3. A metric ton equals 1,000 kilograms or 2,200 pounds.

4. Anne MacLachlan, "Eurodif Amortization 'Challenges' Cost of MOX Fuel for EDF," *NuclearFuel*, January 22, 2001, p. 4.

5. Paul Leventhal and Steven Dolley, "A Japanese Strategic Uranium Reserve: A Safe and Economic Alternative to Plutonium," Nuclear Control Institute, Washington, D.C., January 1994, published in *Science & Global Security* 5 (1994): 1–31; British Nuclear Fuels Ltd. response, "Should Japan Reprocess or Build a Strategic U Reserve?" *Nuclear Engineering International* (April 1994): 28–29; and Steven Dolley, Nuclear Control Institute rebuttal, "Japanese Strategic Uranium Reserve: A Response to BNFL," *Nuclear Engineering International* (September 1994): 50–51.

6. J. Carson Mark, "Explosive Properties of Reactor Grade Plutonium," Nuclear Control Institute, Washington, D.C., 1990, adapted as an article in *Science & Global Security* 4 (1993): 111–28.

7. Letter to Nuclear Control Institute from Hans Blix, director general of the International Atomic Energy Agency, November 1, 1990; and "Blix Says IAEA Does Not Dispute Utility of Reactor-Grade Pu for Weapons," *NuclearFuel*, November 12, 1990, p. 8.

8. J. Carson Mark, Theodore Taylor, Eugene Eyster, William Maraman, and Jacob Wechsler, "Can Terrorists Build Nuclear Weapons?" in *Preventing Nuclear Terrorism*, edited by Paul Leventhal and Yonah Alexander (Lexington, Mass.: Lexington Books, 1987), pp. 55–65.

9. Barbara W. Tuchman, *The March of Folly* (New York: Alfred A. Knopf, 1984), p. 33.

10. Ibid., p. 5.

Chapter 2: Rapporteur's Summary of the Nuclear Control Institute Twentieth Anniversary Conference

1. The list of registrants can be found at the Nuclear Control Institute website http://www.nci.org/conference.htm for the conference.

2. A metric ton equals 1,000 kilograms or 2,200 pounds.

Chapter 4: Nuclear Power and Proliferation

1. "The Need for Nuclear Power," an essay the author wrote with Los Alamos National Laboratory nuclear engineer Denis Beller, provides a detailed technical argument for including nuclear power in the U.S. and world energy future (*Foreign Affairs* 79 [1] [January/February 2000]).

2. Some of the numbers used here come from the writings of Amory Lovins, an articulate and prolific energy theorist.

3. Amory Lovins, "Profiting from a Nuclear-Free Third Millennium," *Power Economics* (November 1999).

4. Arnulf Grübler, Nebojsa Nakicenovic, and David G. Victor, "Dynamics of Energy Technologies and Global Change," *Energy Policy* 27 (1999): 265.

5. Kenneth T. Moore in *Energy and National Security in the 21st Century*, edited by Patrick L. Clawson (Washington, D.C.: National Defense University Press, 1995).

6. Robert L. Bradley Jr., "Renewable Energy," *Cato Policy Analysis* (280)(August 27, 1997): 1.

7. Organisation for Economic Co-operation and Development, Nuclear Energy Agency, "Technical Appraisal of the Current Situation in the Field of Waste Management: A Collective Opinion by the Reactor Waste Management Committee," Paris, 1985.

8. As quoted in Gordon Sims, *The Anti-Nuclear Game* (Ottawa: University of Ottawa Press, 1990).

9. Amory B. Lovins and John H. Price, *Non-Nuclear Futures: The Case for an Ethical Energy Strategy* (Cambridge, Mass.: Ballinger, 1975).

10. Alan D. Pasternak, "Global Energy Futures and Human Development: A Framework for Analysis," UCRL-ID-140773, Lawrence Livermore National Laboratory, Livermore, Calif., 2001.

11. Board of Consultants to the Secretary of State's Committee on Atomic Energy, *A Report on the International Control of Atomic Energy*, U.S. Department of State Publication 2498 (Washington, D.C.: U.S. Government Printing Office, 1946).

12. As quoted in Richard Rhodes, *Dark Sun* (New York: Simon & Schuster, 1995), p. 232.

13. Richard L. Wagner Jr., Edward D. Arthur, and Paul T. Cunningham, "Future Nuclear Energy: Ensuring a U.S. Place at the International Table," unpublished draft, Los Alamos National Laboratory, Los Alamos, N. Mex., August 4, 1997, p. 8.

14. Ibid.

15. Victor Weisskopf's paraphrase, as quoted in Richard Rhodes, *The Making of the Atomic Bomb* (New York: Simon & Schuster, 1986), p. 525.

16. Arnulf Grübler, Nebojsa Nakicenovic, and David G. Victor, "Dynamics of Energy Technologies and Global Change,"*Energy Policy* 27 (1999): 265.

17. Theodore Modis, *Predictions* (New York: Simon & Schuster, 1992), p. 140.

18. "Preface," in *Environmentalists for Nuclear Energy*, Bruno Comby, 2001. Available at http://www.ecolo.org/base/baseen.htm.

19. Ibid.

Chapter 5: Why Nuclear Power's Failure in the Marketplace Is Irreversible (Fortunately for Nonproliferation and Climate Protection)

1. Amory B. Lovins, "Energy Strategy: The Road Not Taken?" *Foreign Affairs* (October 1976).

2. Amory B. Lovins and L. Hunter Lovins, "Fool's Gold in Alaska," *Foreign Affairs* (July/August 2001). Annotated at www.rmi.org/images/other/E-FoolsGold Annotated.pdf.

3. E SOURCE (Boulder, Colo.), semiannual *Electronic Encyclopedia*, www.esource.com.

4. Amory B. Lovins and L. Hunter Lovins, "Climate: Making Sense *and* Making Money," Rocky Mountain Institute, Snowmass, Colo., 1997–1998. Available at www.rmi.org/images/other/C-ClimateMSMM.pdf.

5. William D. Browning and Joseph J. Romm, "Greening the Building and the Bottom Line," Rocky Mountain Institute, Snowmass, Colo., 1998. Available at www.rmi.org/images/other/GDS-GBBL.pdf.

6. Richard Rhodes and Denis Beller, "The Need for Nuclear Power," a paper prepared for presentation at the Nuclear Control Institute's Twentieth Anniversary Conference, "Nuclear Power and the Spread of Nuclear Weapons: Can We Have One without the Other?" Washington, D.C., April 9, 2001. Available at http://www.nci.org/conf/rhodes/index.htm.

7. Steve Nadel, "Lessons Learned," New York State Energy Research and Development Authority, New York State Energy Office, Niagara Mohawk Power Corp., and American Council for an Energy-Efficient Economy, Rept. 90–8, Albany, N.Y., 1990.

8. Paul Hawken, Amory B. Lovins, and L. Hunter Lovins, *Natural Capital-*

ism: Creating the Next Industrial Revolution (New York: Little, Brown, 1999), chapter 6.

9. Amory B. Lovins, "The Super-Efficient Passive Building Frontier," *ASHRAE Journal* (June 1995).

10. Described in a technical book, *Small Is Profitable*, that the author is completing with Rocky Mountain Institute coauthors for publication by Rocky Mountain Institute (www.rmi.org) in early 2002.

11. Amory B. Lovins and Brett D. Williams, "A Strategy for the Hydrogen Transition," National Hydrogen Association, Vienna, Va., April 1999. Available at www.rmi.org/images/other/HC-StrategyHCTrans.pdf.

12. Zeng Peiyan (director, State Development Planning Commission), press statement dated March 6, 2000, reported in *Zhongguo Dianli Bao* (China Electric Power Daily), March 9, 2000.

13. Amory B. Lovins, L. Hunter Lovins, and Leonard Ross, "Nuclear Power and Nuclear Bombs," *Foreign Affairs* (summer 1980).

Chapter 6: Nuclear and Alternative Energy Supply Options for an Environmentally Constrained World: A Long-term Perspective

The author thanks the Geraldine R. Dodge Foundation, the Energy Foundation, the W. Alton Jones Foundation, and the David and Lucile Packard Foundation for support in the preparation of this chapter.

1. Robert H. Williams, "Advanced Energy Supply Technologies," in *Energy and the Challenge of Sustainability*, World Energy Assessment (New York: Bureau for Development Policy, UN Development Programme, 2000), chapter 8, pp. 273–329. This study was sponsored jointly by the UN Development Programme, UN Department of Social and Economic Affairs, and World Energy Council.

2. Ibid., World Energy Assessment, *Energy and the Challenge of Sustainability*. The author was a member of the editorial board for the World Energy Assessment and the convening lead author of chapter 8.

3. A. Grübler, *Technology and Global Change* (Cambridge: Cambridge University Press, 1998).

4. Intergovernmental Panel on Climate Change, *Climate Change 1994—Radiative Forcing of Climate Change and an Evaluation of the IPCC IS92 Scenarios* (Cambridge: Cambridge University Press, 1994).

5. A tonne is a metric ton (1,000 kilograms).

6. J. P. Holdren and K. R. Smith, "Energy, the Environment, and Health," in *Energy and the Challenge of Sustainability*, World Energy Assessment, chapter 3, pp. 61–110.

7. T. M. L. Wigley, R. Richels, and J. A. Edmonds, "Economic and Environmental Choices in the Stabilization of Atmospheric CO_2 Concentration," *Nature* 379 (January 18, 1996): 240–43; and M. I. Hoffert, K. Caldeira, A. K. Jain,

E. F. Haites, L. D. D. Harvey, S. D. Potter, M. E. Schlessinger, S. H. Schneider, R. G. Watts, T. M. L. Wigley, and D. J. Wuebbles, "Energy Implications of Future Stabilization of Atmospheric CO_2 Content," *Nature* 329 (October 29, 1998): 881–84.

8. Thirty-nine percent nuclear, 16 percent hydroelectric, and 45 percent new renewables.

9. Forty-five percent nuclear, 51 percent hydroelectric, and 4 percent new renewables.

10. Both net new and replacement capacity, assuming linear capacity growth and forty-year plant lives.

11. A. I. Filin, V. V. Orlov, V. N. Leonov, A. G. Sila-Novitskij, V. S. Smirnov, and V. S. Tsikunov, "Design Features of BREST Reactors; Experimental Work to Advance the Concept of BREST Reactors; Results and Plans"; R. N. Hill, J. E. Cahalan, H. S. Khalil, and D. C. Wade, "Development of Small, Fast Reactor Core Designs Using Lead-Based Coolant"; A.V. Lopatkin and V.V. Orlov, "Fuel Cycle of BREST 1200 with Non-Proliferation of Plutonium and Equivalent Disposal of Radioactive Waste"; V. Orlov, V. Leonov, A. Sila-Novitski, V. Smirnov, V. Tsikunov, and A. Filin, "Nuclear Power of the Coming Century and Requirements of the Nuclear Technology"; and A.V. Zrodnikov, V. I. Chitaykin, B. F. Gromov, G. I. Toshinsky, U. G. Dragunov, and V. S. Stepanov, "Application of Reactors Cooled by Lead-Bismuth Alloy in Nuclear Power Energy." These papers were presented at Global '99: Nuclear Technology—Bridging the Millennia, International Conference on Future Nuclear Systems, Jackson Hole, Wyo., August 29–September 3, 1999.

12. Williams, "Advanced Energy Supply Technologies."

13. The reactor core volume would be less than 7 cubic meters.

14. Hill et al., "Development of Small, Fast Reactor Core Designs Using Lead-Based Coolant."

15. The deployment rate would be: 408 reactors per year from 2040 to 2055; 816 per year (408 net new plus 408 refurbished) from 2055 to 2070; 1,224 per year (408 net new plus 816 refurbished) from 2070 to 2085; and 1,632 per year (408 net new plus 1,224 per year refurbished) from 2085 to 2100.

16. G. Charpak and R. L. Garwin, *Feux Follets et Champignons Nucleaires* (Paris: Editions Odille Jacob, 1998); and H. Nobukawa et al., "Development of a Floating Type System for Uranium Extraction from Seawater Using Sea Current and Wave Power," *Proceedings of the 4th International Offshore and Polar Engineering Conference*, Osaka, Japan, April 10–15, 1994, pp. 294–300.

17. Operated on a once-through fuel cycle, this reactor requires 8-percent-enriched uranium; the projected fuel burnup is 80,000 megawatt-days per tonne.

18. Harold A. Feiveson, "Diversion-Resistance Criteria for Future Nuclear Power," a paper prepared for the Nuclear Energy and Climate Change Workshop, Center for International Security and Cooperation, Stanford University, Stanford, Calif., June 23, 2000.

19. Ibid.

20. Williams, "Advanced Energy Supply Technologies."

21. Energy R&D Panel of the President's Committee of Advisors on Science and Technology, "Federal Energy Research & Development for the Challenges of the 21st Century," Office of Science and Technology Policy, Executive Office of the President, Washington, D.C., November 1997. Available at http://www.whitehouse.gov/WH/EOP/OSTP/html/ISTP_Home.html.

22. H.-H. Rogner, "Energy Resources," in *Energy and the Challenge of Sustainability*, World Energy Assessment, chapter 5, pp. 135–71.

23. Ibid.

24. H. Kelly and C. H. Weinberg, "Utility Strategies for Using Renewables," in *Renewable Energy: Sources for Fuels and Electricity*, edited by Thomas B. Johansson, Henry Kelly, Amulya K. N. Reddy, and Robert H. Williams (Washington, D.C.: Island Press, 1993), chapter 23, pp. 1011–69.

25. R. B. Shainker, B. Mehta, and R. Pollack, "Overview of CAES [Compressed Air Energy Storage] Technology," *Proceedings of the American Power Conference* (Chicago: Illinois Institute of Technology, 1993): 992–97.

26. Compressed air energy storage costs are low because they are dominated by the turbomachinery components, whose costs are low since they involve gas turbine technology for which the compressor and expander functions are separated in real time.

27. Electric Power Research Institute, "Technical Assessment Guide: Electricity Supply—1993," EPRI TR-102275-V1R7, Volume 1, Rev. 1, Palo Alto, Calif., June 1993.

28. W. Turkenburg, "Renewable Energy Technologies," in *Energy and the Challenge of Sustainability*, World Energy Assessment, chapter 7, pp. 219–72.

29. M. Grubb and N. I. Meyer, "Wind Energy: Resources, Systems, and Regional Strategies," in *Renewable Energy: Sources for Fuels and Electricity*, chapter 4, pp. 157–212.

30. World Energy Council, *New Renewable Energy Resources: A Guide to the Future* (London: Kogan Page, 1994).

31. The author estimated that the global wind energy potential that can be exploited practically is 43,000 terawatt-hours per year. The derivation is as follows. First, the estimate is restricted to wind resources at least as good as Class 4 (an average wind speed of at least 5.6 meters per second 10 meters above the ground—see table 3). Second, the exploitable potential is estimated for a hub height of 100 meters (expected to be a typical hub height with vintage 2030 technology). At this hub height, the estimated net annual rate of electricity generation with 2030 technology is 1,412, 1,566, and 1,797 kilowatt-hours per square meter of area intercepted by the rotor for wind Classes 4, 5, and 6, respectively (see note d, table 3). It is assumed that the spacing of turbines is 10 rotor diameters downwind and 5 rotor diameters across the wind, so that the power density per unit of ground area is (π/200) $\times$ (power density of the wind incident on the turbines), which amounts to 22.2, 24.6, and 28.2 kilowatt-hours per square

meter of ground area for Classes 4, 5, and 6, respectively. Grubb and Meyer ("Wind Energy: Resources, Systems, and Regional Strategies") estimated that globally the land areas over which Class 4 and Classes 5 and higher winds are available are 9.55 million square kilometers and 8.35 million square kilometers, respectively, a total of 14 percent of the land areas of the inhabited continents. The author adopts these estimates but assumes that, for the world, the breakdown between Class 5 and Classes 6 and higher is the same as that for the United States (61.5 percent for Class 5). Under these assumptions the unrestricted global potential is 429,000 terawatt-hours per year (212,000 for Class 4 plus 126,000 for Class 5 plus 91,000 for Classes 6 and higher). Following Grubb and Meyer, the author assumes that the practically exploitable potential is 10 percent of the unrestricted potential. For comparison, the practically exploitable potential for the United States in wind resource Classes 4 and above has been estimated to be 65 percent of the unrestricted potential—assuming that 100 percent of wilderness, 100 percent of urban areas, 50 percent of forests, 30 percent of farmlands, and 10 percent of barren and rangelands are excluded from wind power development (see D. L. Elliott, L. L. Wendell, and G. L. Gower, "An Assessment of the Available Windy Land Area and Wind Energy Potential in the Contiguous United States," Pacific Northwest Laboratories Report PNL-7789, Pacific Northwest Laboratory, Richland, Wash., 1991).

32. This calculation, based to a large extent on A. J. Cavallo ("High-Capacity Factor Wind Energy Systems," *Journal of Solar Energy Engineering* 117 [1995]: 137–43), is for a 6-gigawatt wind farm (with a 36 percent average capacity factor) coupled to a 1.4-gigawatt compressed air energy storage unit operated at a 28 percent capacity factor, with 20 hours of storage, an electricity input/output ratio of 0.67, and a heat rate of 4,326 kilojoules per kilowatt-hour for the expander. The system provides 2 gigawatts of baseload power (90 percent capacity factor). Of this baseload power, 78 percent is provided directly from the wind farm (at 3.7 cents per kilowatt-hour for Class 4 winds in 2020—see table 4) and 22 percent from the compressed air energy storage unit (at 6.7 cents per kilowatt-hour), so that the average generation cost is 4.4 cents per kilowatt-hour and the average heat rate for the wind/compressed air energy storage system is 943 kilojoules per kilowatt-hour. It is assumed that the expander is fired with natural gas, so that the carbon dioxide emission rate for this system is 13 grams of carbon per kilowatt-hour—some 14 percent of the emission rate for a natural gas combined cycle power plant (see table 5).

33. Grubb and Meyer, "Wind Energy: Resources, Systems, and Regional Strategies." For example, in the United States over 95 percent of the high-quality wind resources are concentrated in the twelve states of the Great Plains (see table 3), which account for only 16 percent of the U.S. population and whose land areas (34 percent of total U.S. land) are occupied mainly by ranchers and farmers. In China, excellent wind resources are available on 83,000 square kilometers (0.9 percent of China's land area) in sparsely populated Inner Mongolia, where the

potential production is 1,800 terawatts-hours per year (D. Lew, R. H. Williams, S. Xie, and S. Zhang, "Large-Scale Baseload Wind Power in China," *Natural Resources Forum* 22 [3][1998]: 165–218)—about twice the rate of thermal electricity generation rate in China in 1998.

34. Cavallo, "High Capacity Factor Wind Energy Systems," and ibid., Lew et al.

35. For Class 4 winds, U.S. wind power costs are projected to average 4.3 cents per kilowatt-hour in 2005 (see table 4), when the net rate of generation per unit area of wind intercepted is expected to be about 1,300 kilowatt-hours per square meter per year for 70-meter hub heights (see table 4). For a wind turbine spacing of 5 turbine rotor diameters across the wind and 10 diameters downwind, the corresponding rate of generation per unit of ground area is $(\pi/200) \times$ 1,300, or 20 kilowatt-hours per square meter per year. Assuming that the royalty rate to the landowner is 2.5 percent of revenues generated (see table 4), the royalty amounts to $88 per acre per year. For comparison, net U.S. farm income in 1999 was $51 per acre, of which $24 per acre were direct government payments. (The farm income per acre was calculated as follows. Total U.S. farm income in 1999 was $48 billion, of which $23 billion were direct government payments [M. Morehart, J. Ryan, D. Peacock, and R. Strickland, "U.S. Farm Income Decline in 2000 To Be Tempered by Government Payments," *Agricultural Outlook*, Economic Research Service, U.S. Department of Agriculture {January–February, 2000}]; and the total U.S. farm area in 1999 was 947 million acres [R. J. Shapiro, *Statistical Abstract of the United States*, 120th ed. {Washington, D.C.: Bureau of the Census, U.S. Department of Commerce}].)

36. A. Cabraal, M. Cosgrove-Davies, and L. Schaeffer, "Best Practices for Photovoltaic Household Electrification Programs: Lessons from Experiences in Selected Countries," Technical Paper No. 324, World Bank, Washington, D.C., 1996.

37. Energy R&D Panel, "Federal Energy Research & Development. . . ."

38. Net metering is a policy that allows customers to run their electric meters backwards, delivering excess electricity to the grid for credit at retail rates during periods when photovoltaic generation exceeds on-site demand. In the United States, thirty states have adopted net metering policies to encourage the deployment of photovoltaic systems.

39. A. Payne, R. Duke, and R. Williams, "Accelerating Residential PV Expansion: Supply Analysis for Competitive Electricity Markets," *Energy Policy* 29 (2001): 787–800.

40. C. Marnay, R. C. Richey, S. A. Mahler, and R. J. Markel, "Estimating the Environmental and Economic Effects of Widespread Residential PV Adoption Using GIS and NEMS," a paper presented at the 1997 American Solar Energy Society Meeting, Washington, D.C., May 1997.

41. Energy payback is the time required to pay back the energy invested in the manufacturing and installation of photovoltaic systems. It must be a small

fraction of system lifetimes (~twenty to twenty-five years) if photovoltaics are to make substantial contributions to energy supplies. For current grid-connected rooftop crystalline silicon and thin-film systems, the payback times are four to nine years and three to four years, respectively. The payback times are projected to fall in less than a decade's time to three to four years and one to two years for crystalline silicon and thin-film photovoltaic systems, respectively (E. A. Alsema, P. Frankl, and K. Kato, "Energy Payback Time of Photovoltaic Energy Systems: Present Status and Prospects," in *Proceedings of the Second World Conference and Exhibition on PV Solar Energy Conversion*, edited by J. Schmid et al., Report EUR 188656 EN [Brussels: 1998]; and Turkenburg, "Renewable Energy Technologies").

42. Electric Power Research Institute and the Office of Utility Technologies, Energy Efficiency and Renewable Energy, U.S. Department of Energy, "Renewable Energy Technology Characterizations," EPRI TR-10949, Electric Power Research Institute, Palo Alto, Calif., December 1997. Very low costs for thin-film photovoltaic technologies are potentially realizable in large part because the layer of active photovoltaic material (deposited on a glass, steel, or other substrate) is typically only about 1 micron thick (about 1 percent of the thickness of a human hair), so that overall costs for the active photovoltaic materials are low.

43. Assuming an overall photovoltaic efficiency of 13.6 percent, the average efficiency projected for thin-film photovoltaic systems in 2030 (Electric Power Research Institute and the Office of Utility Technologies, "Renewable Energy Technology Characterizations"), the peak photovoltaic output is 136 watts per square meter (because the peak insolation is 1,000 watts per square meter). The required collector area for a 1-kilowatt (electric) system is thus (1,000 watts)/(136 watts per square meter), or 7.35 square meters.

44. However, the power sector would not necessarily be free of fossil energy in such a renewables-intensive electricity future if compressed air energy storage units powered by fossil fuels were to be used to enable intermittent renewables to follow load or provide baseload power. Suppose that natural gas is used as the fuel in compressed air energy storage units to back up *all* wind and photovoltaic power at a rate of 1,000 kilojoules per kilowatt-hour (see note 32). Assuming that in 2100 wind and photovoltaic power are produced at rates of 43,000 and 16,000 terawatt-hours per year, respectively, the total natural gas required in 2100 would be 59 exajoules per year and the corresponding carbon dioxide emission rate would be 0.80 gigatonne of carbon per year. Alternatively, compressed air energy storage units could be powered by hydrogen derived from fossil fuels with sequestration of the carbon dioxide coproduct, in which case the carbon dioxide emissions from the compressed air energy storage units would be negligible.

45. Assuming insolation of 1,800 kilowatt-hours per square meter per year (the average for the United States), a 0.8-kilowatt (electric) array would produce 1,440 kilowatt-hours of electricity per square meter per year. Assuming deploy-

ment of photovoltaic capacity at an average rate of 0.8 kilowatts per capita in 2100 and a world population of 11.3 billion, the worldwide photovoltaic generation rate in 2100 would be 16,000 terawatt-hours per year.

46. Williams, "Advanced Energy Supply Technologies"; and S. Bachu, "Geological Sequestration of Anthropogenic Carbon Dioxide: Applicability and Current Issues," in *Geological Perspectives of Global Climate Change,* edited by L. C. Gerhard, W. E. Harrison, and B. M. Hanson, AAPG Studies in Geology 47 (Tulsa, Okla.: American Association of Petroleum Geologists, 2001), pp. 285–303.

47. Ibid., Williams.

48. Most enhanced oil recovery projects in the United States are in the Permian Basin of Texas. Most of the carbon dioxide for these projects is transported by pipeline from natural reservoirs of carbon dioxide in Colorado, New Mexico, and Wyoming (for example, via an 800-kilometer pipeline from the McElmo Dome in western Colorado, which contains 0.5 gigatonne of carbon dioxide).

49. S. H. Stevens, V. A. Kuuskraa, and J. Gale, "Sequestration of Carbon Dioxide in Depleted Oil and Gas Fields: Global Capacity, Costs, and Barriers," in *Greenhouse Gas Control Technologies: Proceedings of the 5th International Conference on GHG [Greenhouse Gas] Control Technologies (GHGT-5),* edited by D. J. Williams, R. A. Durie, P. McMullan, C. A. J. Paulson, and A. Y. Smith (Collingwood, Victoria, Australia: CSIRO Publishing, 2000), pp. 278–83.

50. S. H. Stevens, V. A. Kuuskraa, D. Spector, and P. Riemer, "Enhanced Coalbed Methane Recovery Using Carbon Dioxide Injection: Worldwide Resource and Carbon Dioxide Injection Potential," *Greenhouse Gas Control Technologies: Proceedings of the 4th International Conference on GHG Control Technologies,* edited by B. Eliasson, P. Riemer, and A. Wokaun (Amsterdam: Pergamon, 1999), pp. 175–80.

51. W. D. Gunter, T. Gentzix, B. A. Rottenfusser, and R. J. H. Richardson, "Deep Coalbed Methane in Alberta, Canada: A Fuel Resource with the Potential of Zero Greenhouse Emissions," *Energy Conversion and Management* 38 (1997): S217–22.

52. Ibid.

53. Of the 6 trillion tonnes of U.S. coal resources at depths up to 1,800 meters, 90 percent cannot be mined with current technology because the coal is too deep, the seams are too thin, or the mining would be unsafe (C. W. Byrer and H. D. Guthrie, "Coal Deposits: Potential Geological Sink for Sequestering Carbon Dioxide Emissions from Power Plants," in *Greenhouse Gas Control Technologies . . .* , pp. 181–87).

54. Stevens, Kuuskraa, and Gale, "Sequestration of Carbon Dioxide in Depleted Oil and Gas Fields. . . ."

55. M. J. van der Burgt, J. Cantle, and V. K. Boutkan, "Carbon Dioxide Disposal from Coal-Based IGCCs [Integrated Gasifier/Combined Cycles] in Depleted Gas Fields," *Energy Conversion and Management* 33 (5–8)(1992): 603–10;

and I. R. Summerfield, S. H. Goldthorpe, N. Williams, and A. Sheikh, "Costs of Carbon Dioxide Disposal Options," in *Proceedings of the International Energy Agency Carbon Dioxide Disposal Symposium* (Amsterdam: Pergamon, 1993).

56. Capacity from past production plus proved reserves plus estimated undiscovered conventional resources.

57. C. A. Hendriks, "Carbon Dioxide Removal from Coal-Fired Power Plants," Ph.D. thesis, Department of Science, Technology, and Society, Utrecht University, Utrecht, The Netherlands, 1994.

58. Intergovernmental Panel on Climate Change, "Energy Supply Mitigation Options," in *Climate Change 1995: Impacts, Adaptations and Mitigation of Climate Change: Scientific-Technical Analyses*, Second Assessment Report of the Intergovernmental Panel on Climate Change (Cambridge: Cambridge University Press, 1996).

59. B. Hitchon, W. D. Gunter, T. Gentzis, and R. Bailey, "Sedimentary Basins and Greenhouse Gases: A Serendipitous Association," *Energy Conversion and Management* 40 (1999): 825–43; S. Bachu and W. D.Gunter, "Storage Capacity of Carbon Dioxide in Geological Media in Sedimentary Basins with Application to the Alberta Basin," in *Greenhouse Gas Control Technologies . . .*, pp. 195–200; and Bachu, "Geological Sequestration of Anthropogenic Carbon Dioxide. . . ."

60. The critical point for carbon dioxide is 74 atmospheres and 31°C.

61. The hydrostatic pressure gradient is typically about 100 atmospheres per kilometer.

62. Deep aquifers (~800 meters or more below the surface) tend to be saline because the contained water is fossil water that has been there over sufficient geological time for the water to come into chemical equilibrium with the minerals in the host rock. Dissolved salts typically make the water brackish and often even briny.

63. Hendriks, "Carbon Dioxide Removal from Coal-Fired Power Plants."

64. S. Bachu, W. D. Gunter, and E. H. Perkins, "Aquifer Disposal of CO_2: Hydrodynamic and Mineral Trapping," *Energy Conversion and Management* 35 (1994): 269–79; and S. Holloway, ed., "The Underground Storage of Carbon Dioxide," a report prepared for the Joule II Programme (DG XII) of the Commission of the European Communities, Contract No. JOU2 CT92–0031, Brussels, Belgium, February 1996.

65. Ibid., Bachu, Gunter, and Perkins.

66. W. D. Gunter, E. H. Perkins, and T. J. McCann, "Aquifer Disposal of Carbon Dioxide-Rich Gases: Reaction Design for Added Capacity," *Energy Conversion and Management* 34 (1993): 941–48.

67. W. Ormerod, "The Disposal of Carbon Dioxide from Fossil Fuel Power Stations," IEA/GHG/SR3, IEA Greenhouse Gas Research and Development Programme, International Energy Agency, Cheltenham, Eng., 1994.

68. Hendriks, "Carbon Dioxide Removal from Coal-Fired Power Plants."

69. Rogner, "Energy Resources."

70. O. Kaarstad, "Emission-Free Fossil Energy from Norway," *Energy Conversion and Management* 33 (5–8)(1992): 781–86.

71. International Energy Agency, "Carbon Dioxide Capture and Storage in the Natuna NG [Natural Gas] Project," *Greenhouse Issues* 22 (1996): 1.

72. H. L. Longworth, G. C. Dunn, and M. Semchuck, "Underground Disposal of Acid Gas in Alberta, Canada: Regulatory Concerns and Case Histories," SPE 35584, a paper presented at the Gas Technology Conference, Calgary, Alberta, Canada, April 28–May 1, 1996; and E. Wichert and T. Royan, "Acid Gas Injection Eliminates Sulfur Recovery Expense," *Oil and Gas Journal* (April 28, 1997): 67–72.

73. Holloway, "The Underground Storage of Carbon Dioxide"; R. H. Socolow, ed., "Fuels Decarbonization and Carbon Sequestration: Report of a Workshop by the Members of the Report Committee," PU/CEES Report 302, Center for Energy and Environmental Studies, Princeton University, Princeton, N.J., 1997, available at http://www.princeton.edu/~ceesdoe; and Energy R&D Panel, "Federal Energy Research & Development. . . ."

74. S. Holloway, "Safety of Underground Disposal of Carbon Dioxide," *Energy Conversion and Management* 38 (1997): S241–45.

75. W. D. Gunter, R. J. Chalaturnyk, and J. D. Scott, "Monitoring of Aquifer Disposal of Carbon Dioxide: Experience from Underground Gas Storage and Enhanced Oil Recovery," in *Greenhouse Gas Control Technologies* . . . , pp. 151–56.

76. Williams, "Advanced Energy Supply Technologies."

77. Ibid.

78. Ibid.

79. The installed capital and operation and maintenance costs for a new 35.5-percent-efficient subcritical coal steam-electric plant with flue gas desulfurization are $1,090 per kilowatt (electric) and 0.43 cents per kilowatt-hour, respectively (Williams, "Advanced Energy Supply Technologies"), so that for a coal price of $0.93 per gigajoule, the generation cost is 3.70 cents per kilowatt-hour, assuming an annual capital charge rate of 15 percent and a capacity factor of 80 percent. For comparison, the estimated generation cost for a coal integrated gasifier/combined cycle plant is 3.75 cents per kilowatt-hour (see table 5).

80. At such high partial pressures, use of a physical solvent such as Selexol (dimethyl ether of polyethylene glycol) for carbon dioxide removal leads to a lower lifecycle cost than for carbon dioxide removal using amines.

81. The overnight construction cost of the advanced boiling water reactor is estimated to be $1,582 per kilowatt (electric) (Nuclear Energy Agency, "Reductions of Capital Costs of Nuclear Power Plants," Organisation for Economic Cooperation and Development, Paris, France, 2000). Assuming a 5-year construction period and a 10 percent real interest rate during construction, the total installed cost is $1,932 per kilowatt, so that the annual capital charge with an 80

percent capacity factor is 4.14 cents per kilowatt-hour. Assuming a nuclear fuel cost of 0.54 cents per kilowatt-hour (see table 5), the operation and maintenance cost for the advanced boiling water reactor would have to be less than 1.0 cent per kilowatt-hour to compete with the decarbonized coal integrated gasifier/combined cycle power plant (see table 5).

82. J. M. Ogden, and R. H. Williams, *Solar Hydrogen: Moving Beyond Fossil Fuels* (Washington, D.C.: World Resources Institute, 1989).

83. J. M. Ogden, R. H. Williams, and E. D. Larson, "Toward a Hydrogen-Based Transportation System," draft manuscript, Princeton Environmental Institute, Princeton University, Princeton, N.J., May 2001.

84. Typically the manufacture of hydrogen from natural gas begins by making steam react with the natural gas at a temperature of ~900°C to make synthesis gas (mostly carbon monoxide and hydrogen). The synthesis gas is then cooled, and the carbon monoxide is reacted with more steam to produce hydrogen and carbon dioxide via the water-gas-shift reaction, so that the overall processing leads to a gaseous mixture consisting mainly of hydrogen and carbon dioxide. The hydrogen is next separated from the carbon dioxide and other gases. Commercial technology can provide hydrogen that is 99.999 percent pure. If a stream of relatively pure carbon dioxide is desired as a coproduct (to facilitate carbon dioxide disposal), the cost is somewhat higher than that for the standard practice.

85. Robert H. Williams, "Toward Zero Emissions for Transportation Using Fossil Fuels," a paper prepared for Managing Transitions in the Transportation Sector: How Fast and How Far? VIII Biennial Conference on Transportation, Energy, and Environmental Policy, Asilomar Conference Center, Monterey, Calif., September 11–14, 2001. The Transportation Research Board will publish the conference proceedings in 2002.

86. Ibid.; and Robert H. Williams, "Toward Zero Emissions for Coal: Roles for Inorganic Membranes," *Proceedings of the International Symposium Toward Zero Emissions: The Challenge for Hydrocarbons* (Rome: EniTecnologie, March 1999): 212–42.

87. Williams, "Toward Zero Emissions for Transportation Using Fossil Fuels."

88. Ogden, Williams, and Larson, "Toward a Hydrogen-Based Transportation System."

89. It is assumed that by 2100 the only fossil fuels used directly without carbon dioxide capture and sequestration would be coal for iron and steelmaking and jet fuel for airplane use.

It is assumed that by 2100 iron and steel would be made using smelt reduction/near final shape casting (an advanced process now under development that is widely viewed as the technology of choice for the future), with the current mix of iron ore and scrap specified by the Organisation for Economic Co-operation and Development, so that the coal required is 5.9 gigajoules per tonne of steel (E. Worrell, "Advanced Technologies and Energy Efficiency in the Iron and Steel

Industry in China," *Energy for Sustainable Development* 2 [4][November 1995]: 27–40). It is further assumed that the average per capita rate of steel consumption for a global population of 11.3 billion is 386 kilograms (the average rate the Organisation for Economic Co-operation and Development gives for 1987). Thus, coal use for iron and steelmaking is 25.8 exajoules per year. Assuming a carbon dioxide emission rate for coal of 25.3 kilograms of carbon per gigajoule, total emissions from iron and steelmaking in 2100 would be 0.65 gigatonnes of carbon per year.

It is assumed that the global requirement for jet fuel for airplane use in 2100 is 58.8 exajoules per year, calculated as: U.S. jet fuel in 1998 times global gross domestic product in 2100 divided by U.S. gross domestic product in 1998 divided by $(1.007)^{102}$. The rate of U.S. jet fuel consumption in 1998 is 3.59 exajoules (Energy Information Administration, *Annual Energy Outlook 2001: With Projections to 2020,* DOE/EIA-0383 [2001] [Washington, D.C.: U.S. Department of Energy, December 2000]). In IS92a, global gross domestic product in 2100 is $284.5 trillion (in 1996 dollars). U.S. gross domestic product in 1998 is $8.52 trillion (in 1996 dollars). It is assumed that the efficiency of airplanes increases at an average rate of 0.7 percent per year (the energy efficiency of new production aircraft has improved at a rate of 1–2 percent per year since the dawn of the jet era); a recent panel of experts of the Intergovernmental Panel on Climate Change projected that a 0.7 percent per year rate of improvement can be expected from 1997 to 2050 (J. S. Lewis and R. W. Niedzwiecki, "Aircraft Technology and Its Relation to Emissions," in *Aviation and the Global Atmosphere,* Intergovernmental Panel on Climate Change [Oxford: Oxford University Press, 1999], chapter 7). It is further assumed that refineries are 90 percent efficient in the manufacture of oil products from crude oil so that the total oil required to serve the jet fuel markets in 2100 is 63.3 exajoules per year (58.8 divided by 0.9), and the corresponding carbon dioxide emission rate is 1.16 gigatonnes of carbon per year.

90. For any new energy technology launched from a zero base to be able to make major supply contributions later in this century, it must expand initially at an accelerated pace, with growth rates in the range of 30–40 percent per year for a period on the order of a couple of decades. Such market-launching growth rates characterized nuclear power in its early years: Its growth worldwide averaged 37 percent per year from 1957 to 1977 (Robert H. Williams and G. Terzian, "A Benefit/Cost Analysis of Photovoltaic Technology," PU/CEES Report No. 281, Center for Energy and Environmental Studies, Princeton University, Princeton, N.J., October 1993), and, as noted, wind power has been expanding at rates near 30 percent per year since the early 1990s.

91. It is estimated that for wind power at 2.9 cents per kilowatt-hour (the projected generation cost for Class 6 wind in 2020—see table 4), the cost of baseload wind power (a wind farm plus compressed air energy storage) would be 3.6 cents per kilowatt-hour. Transmitting this electricity 300 kilometers via a 1-

gigawatt high-voltage alternating current line (including ohmic losses) would bring the city-gate cost of this baseload power to 4 cents per kilowatt-hour.

92. The cost of hydrogen in dollars per gigajoule using *advanced* electrolytic technology is:

$$\$3.59 + (\text{electricity price})/(0.882 \times 0.0036 \text{ gigajoule per kilowatt-hour}),$$

where the electricity price is in dollars per kilowatt-hour and 0.882 is the efficiency of converting alternating current into hydrogen (an electrolytic efficiency of 0.90, higher heating value basis, times a rectifier efficiency of 0.98). This estimate is based on an analysis of performance and costs for advanced bipolar electrolyzers in J. M. Ogden and J. Nitsch, "Solar Hydrogen," in *Renewable Energy: Sources for Fuels and Electricity*, chapter 22, pp. 925–1009, assuming a 15 percent annual capital charge rate and an 80 percent capacity factor. Thus, if the price of electricity is \$0.04 per kilowatt-hour, the cost of hydrogen energy is \$16.2 per gigajoule.

93. Assuming a coal price of \$0.93 per gigajoule (see table 5) and a carbon tax of \$100 per tonne, the estimated cost of producing hydrogen from coal with the carbon dioxide sequestered using near-commercial technology is \$6.4 per gigajoule (Williams, "Toward Zero Emissions for Transportation Using Fossil Fuels"). Thus the price of carbon-free electricity (in dollars per kilowatt-hour) at which the cost of electrolytic hydrogen (based on the use of advanced electrolytic technology) equals the cost of hydrogen from coal with the carbon dioxide sequestered is \$0.0089 per kilowatt-hour, the solution of the equation (see note 92)

$$6.40 = \$3.59 + (\text{electricity price})/(0.882 \times 0.0036 \text{ gigajoule per kilowatt-hour}).$$

If, instead, present-day electrolytic technology (73.9 percent efficient and 70 percent more capital-intensive than future technology) were used, the break-even electricity price is \$0.00074 per kilowatt-hour, the solution of the equation (Williams, "Toward Zero Emissions for Transportation Using Fossil Fuels")

$$6.40 = \$6.12 + (\text{electricity price})/(0.739 \times 0.0036 \text{ gigajoule per kilowatt-hour}).$$

94. Note in table 1 that under IS92a in 2100 fossil fuels used directly amount to 541 exajoules per year, compared to 245 exajoules (68,000 terawatt-hours) per year for electricity generation.

95. K. Yoshida, "Present Status of R&D for Hydrogen Production from Water in Japan," *Energy Research* 7 (1983): 1–12; and S. Yalçin, "A Review of Nuclear Hydrogen Production," *International Journal of Hydrogen Energy* 14 (8)(1989): 551–61.

96. A. Steinfeld and R. Palumbo, "Fuels from Sunlight and Water," a paper

available at the website of the Paul Scherrer Institute, Switzerland, 2001, www.psi.ch.

97. Yoshida, "Present Status of R&D for Hydrogen Production from Water in Japan."

98. The UT-3 process is based on the following reactions aimed at decomposing water thermochemically:

$$CaO + Br_2 \rightarrow Ca\,Br_2 + {}^1\!/_2\, O_2 \text{ (at 700–750°C)}$$
$$CaBr_2 + H_2O \rightarrow CaO + HBr \text{ (at 500–600°C)}$$
$$Fe_3O_4 + 8\,HBr \rightarrow 3\,FeBr_2 + H_2O + Br_2 \text{ (at 200–300°C)}$$
$$3\,FeBr_2 + 4\,H_2O \rightarrow Fe_3O_4 + 6\,HBr + H_2 \text{ (at 550–600°C)}$$

where CaO is calcium oxide (lime), Br_2 is bromine, $CaBr_2$ is calcium bromide, O_2 is oxygen, H_2O is water, HBr is hydrobromic acid, Fe_3O_4 is iron oxide (magnetite), $FeBr_2$ is iron bromide, and H_2 is hydrogen.

In this series of reactions, which take place in multiple vessels, water and heat are consumed, and both hydrogen and oxygen are produced; the rest of the chemicals are recycled.

99. Y. Tadokoro, T. Kajiyama, T. Yamaguchi, N. Sakai, H. Kameyama, and K. Yoshida, "Technical Evaluation of UT-3 Thermochemical Hydrogen Production Process for an Industrial Scale Plant," *International Journal of Hydrogen Energy* 22 (1)(1997): 49–56.

100. Williams, "Toward Zero Emissions for Transportation Using Fossil Fuels."

101. Tadokoro, Kajiyama, Yamaguchi, Sakai, Kameyama, and Yoshida, "Technical Evaluation of UT-3 Thermochemical Hydrogen Production Process. . . ."

102. ESKOM, the South African electric utility, is working to develop the pebble-bed modular reactor, a small modular high-temperature gas-cooled reactor (a 1,100-megawatt [electric] power plant would be made up of ten 110-megawatt [electric] modules). The utility is aiming for (hoping for) an installed capital cost of $1,000 per kilowatt (electric). If it attains this capital cost target, the generation cost (assuming an annual capital charge rate of 15 percent and a capacity factor of 80 percent) would be 3 cents per kilowatt-hour (J. T. Taylor, "Economic and Market Potential of Small Innovative Reactors," a paper presented at the Workshop on New Energy Technologies: A Policy Framework for Micro-Nuclear Technology, Houston, Texas, March 19–20, 2001). The breakdown is as follows:

Capital	$0.0215	
Nuclear fuel	0.0047	
Operation and maintenance	0.0030	
Decommissioning	0.0010	
Total	$0.0302	per kilowatt-hour

Based on these cost targets, a rough estimate of the cost of nuclear heat is obtained by assuming that the pebble-bed modular reactor is 43 percent efficient in converting nuclear heat into electricity and that 40 percent of the capital cost for the electricity generation is for power conversion equipment not needed when the pebble-bed modular reactor is used to provide heat for the manufacture of hydrogen. The cost of nuclear heat estimated under these assumptions is $2.58 per gigajoule.

Because the cost goals for the pebble-bed modular reactor are very aggressive and the pebble-bed modular reactor is still in the research and development phase, it is very uncertain if the cost goals are realizable.

103. Intergovernmental Panel on Climate Change, "Energy Supply Mitigation Options."

104. For comparison, the land areas in croplands, forests, and pastures are 15 million, 41 million, and 33 million square kilometers, respectively.

105. Assuming an average productivity of 12 tonnes per hectare per year and a biomass heating value of 20 gigajoules per tonne.

106. Turkenburg, "Renewable Energy Technologies."

107. World Energy Assessment, *Energy and the Challenge of Sustainability.*

108. Turkenburg, "Renewable Energy Technologies."

Chapter 7: A World with, or without, Nuclear Power?

1. Since 1950 the author has worked with the U.S. government to build nuclear weapons and also to control them through the Limited Test Ban Treaty, Comprehensive Nuclear Test Ban Treaty, Nuclear Non-Proliferation Treaty, and more recently the Strategic Arms Limitation Treaty and Strategic Arms Reduction Treaty agreements. For more than twenty years the author has been a member of the Committee on International Security and Arms Control of the National Academy of Sciences, which has exerted major efforts to control and dispose of excess weapon plutonium and uranium.

2. Wolfgang K. H. Panofsky, "Management and Disposition of Excess Weapons Plutonium," report of the Committee on International Security and Arms Control, National Academy of Sciences (Washington, D.C.: National Academy Press, January 1994), pp. 32–33; and U.S. Department of Energy, "Final Nonproliferation and Arms Control Assessment of Weapons-Usable Fissile Material Storage and Excess Plutonium Disposition Alternatives," Washington, D.C., 1997.

3. Ibid., U.S. Department of Energy, pp. 38–39.

4. Ibid.

5. J. Carson Mark, "Explosive Properties of Reactor-Grade Plutonium," *Science & Global Security* 4 (1993): 111–28.

6. Panofsky, "Management and Disposition of Excess Weapons Plutonium."

7. Richard L. Garwin, "Post-Cold War World and Nuclear Weapons Proliferation," a paper delivered at Session 5, "Nuclear Nonproliferation and Plutonium," of the 29th Japan Atomic Industrial Forum (JAIF) Annual Conference, Nagoya, Japan, April 19, 1996.

8. Spurgeon M. Keeny Jr. et al., "Nuclear Power Issues and Choices" (Cambridge: Ballinger Publishing, March 1977); and Harold W. Lewis et al., "Report to the APS [American Physical Society] by the Study Group on Light-Water Reactor Safety," *Reviews of Modern Physics* 47 (Supplement No. 1, June 1975).

9. Richard L. Garwin, letter dated September 1999 commenting on an article by Zbigniew Jaworowski regarding the linear relationship of cancer to dose at low levels of ionizing radiation, in *Physics Today* 53 (5)(May 2000): 12–14.

10. UN Special Committee on the Effects of Atomic Radiation, "Sources and Effects of Ionizing Radiation," Report to the General Assembly, with Scientific Annexes, United Nations, N.Y., 1993.

11. Ibid.

12. Garwin, letter dated September 1999. Those interested can refer to the author's earlier publications on this issue.

13. Director of Safety, Health, and Environment, British Nuclear Fuels Ltd., "Annual Report on Discharges and Monitoring of the Environment," Risley, Warrington, Cheshire, WA3 6AS, U.K., 1997; and letter from Roger Coates, British Nuclear Fuels Ltd. to Richard.L. Garwin, April 6, 1999.

14. UN Special Committee on the Effects of Atomic Radiation, "Sources and Effects of Ionizing Radiation."

15. Edward Teller, "Fast Reactors: Maybe," *Nuclear News* (August 21, 1967).

16. John P. Holdren and Rajendra K. Pachauri, "Energy," in *An Agenda of Science for Environment and Development into the 21st Century,* International Council for Science (Cambridge: Cambridge University Press, 1992), pp. 103–18.

17. Richard L. Garwin, "L'uranium extrait de l'eau de mer: un combustible vert pour demain? [Seawater uranium: a green fuel for the future?]," in "Les Energies du Futur" [Energies of the Future], *Revue des Deux Mondes* (Paris, France) (April 2001): 67–77.

18. T. Kato, K. Okugawa, Y. Sugihara, and T. Matsumura, "Conceptual Design of Uranium Recovery Plant from Seawater," undated (perhaps March 2001).

19. Jacques Foos et al., "Document of 1996 on Extraction of Uranium from Seawater," COGEMA, June 1996.

20. Frederick P. Brooks, *The Mythical Man-Month* (Reading, Mass.: Addison Wesley Longman, 1982).

Chapter 8: Attempts to Reduce the Proliferation Risks of Nuclear Power: Past and Current Initiatives

1. Amory B. Lovins, *Soft Energy Paths: Toward a Durable Peace* (Cambridge: Ballinger Press, 1977), chapter 11.

2. Edward Teller, *Energy From Heaven and Earth* (San Francisco: W. H. Freeman, 1979), pp. 192–93.

3. As quoted by A. W. Weinberg, "From Technological Fixer to Think-Tanker," in *Annual Review of Energy and the Environment* (Palo Alto, Calif.: Annual Reviews, 1994), p. 17.

4. Board of Consultants to the Secretary of State's Committee on Atomic Energy, *A Report on the International Control of Atomic Energy*, U.S. Department of State Publication 2498 (Washington, D.C.: U.S. Government Printing Office, 1946).

5. See especially George Perkovich, *India's Nuclear Bomb* (Berkeley: University of California Press, 1999).

6. J. Carson Mark, "Reactor-Grade Plutonium's Explosive Properties," Nuclear Control Institute, Washington, D.C., August 1980, reprinted in *Science & Global Security* 4 (1993): 111–28.

7. See, for example, Richard L. Garwin, "Reactor-Grade Plutonium Can be Used to Make Powerful and Reliable Nuclear Weapons: Separated Plutonium in the Fuel Cycle Must Be Protected as If It Were Nuclear Weapons," newsletter of the Nuclear Information Service, Japan, August 26, 1998. Available at www.fas.org/rlg.

8. B. Goodwin and J. Kammerdiener, "Future Proliferation Threat," a paper presented at the Proliferation-Resistant Nuclear Power Systems workshop, Center for Global Security Research, Lawrence Livermore National Laboratory, Livermore, Calif., June 2–4, 1999.

9. U.S. Department of Energy, "Nonproliferation and Arms Control Assessment of Weapons-Usable Fissile Material Storage and Disposition Alternatives," draft, Washington, D.C., October 1996, pp. 37–39.

10. U.S. Department of Energy, *Nuclear Proliferation and Civilian Nuclear Power: Report of the Nonproliferation Alternative Systems Assessment Program (NASAP), Volume II: Proliferation Resistance*, DOE/NE-0001/2 (Washington, D.C.: U.S. Government Printing Office, June 1980), pp. 3-33–3-34.

11. See, for example, the articles in *International Arrangements for Nuclear Fuel Reprocessing*, edited by A. Chayes and W. B. Lewis (Cambridge: Ballinger Press, 1977).

12. Y. I. Chang and C. E. Till, "The Integral Fast Reactor," *Advances in Nuclear Science & Technology* 20 (1988): 127–54.

13. A. Galperin, P. Reichert, and A. Radkowsky, "Thorium Fuel for Light

Water Reactors—Reducing the Proliferation Potential of the Nuclear Fuel Cycle," *Science & Global Security* 6 (1997): 265–90.

14. However, it has been demonstrated—at least on paper—that a submarine reactor with an initial fuel enrichment of 20 percent can have the same long core life (twenty years) as a reactor with an initial enrichment of 97.3 percent. The major penalty is that the volume of the core of the 20-percent-enriched design is about three times larger. See T. D. Ippolito Jr., "Effects of Variation of Uranium Enrichment on Nuclear Submarine Reactor Design," master's degree thesis, Department of Nuclear Engineering, Massachusetts Institute of Technology, Cambridge, Mass., May 1990.

15. U.S. Department of Energy, *Nuclear Proliferation and Civilian Nuclear Power . . .* , pp. 2-25 and 2-39.

16. T. Kato et al., "Conceptual Design of Uranium Recovery Plant from Seawater" (in Japanese), *Journal of Thermal and Nuclear Power Engineering Society* 50 (1999): 71–77.

17. See also the recent paper by Harold A. Feiveson, "Comments on the Development Path for Proliferation-Resistant Nuclear Power," presented at a conference at the James Baker Institute, Rice University, Houston, Texas, March 19–20, 2001.

Chapter 9: Technical Opportunities for Increasing Proliferation Resistance of Nuclear Power Systems (TOPS) Task Force

The views presented are those of the author and do not represent those of Lawrence Livermore National Laboratory, the University of California, or the United States government.

1. "Technological Opportunities to Increase the Proliferation Resistance of Global Civilian Nuclear Power Systems (TOPS)," report by the TOPS Task Force of the Nuclear Energy Research Advisory Committee (NERAC), U.S. Department of Energy, Washington, D.C., January 2001, available at http://www.nuclear.gov/nerac/FinalTOPSRpt.pdf; and "Annex: Attributes of Proliferation Resistance for Civilian Nuclear Power Systems," Nuclear Energy Research Advisory Committee (NERAC), U.S. Department of Energy, October 2000, available at http://www.nuclear.gov/nerac/FinalTOPSRptAnnex.pdf.

2. "Proliferation-Resistant Nuclear Power Systems: A Workshop on New Ideas" (held June 2–4, 1999), UCRL-JC-137954 and CGR-2000-001, Center for Global Security Research, Lawrence Livermore National Laboratory, Livermore, Calif., March 2000, available at http://cgsr.llnl.gov/Final_Wkshp_Rpt.pdf; and "Report of the International Workshop on Technology Opportunities for Increasing the Proliferation Resistance of Global Civilian Nuclear Power Systems (TOPS)," held in Washington, D.C., March 29–30, 2000, and sponsored by the Nuclear Energy Research Advisory Committee and the Center for Global Security Research at Lawrence Livermore National Laboratory, Livermore, Calif. Available at www.nuclear.gov/nerac/tops.pdf.

3. U.S. Department of Energy, *Nuclear Proliferation and Civilian Nuclear Power: Report of the Nonproliferation Alternative Systems Assessment Program (NASAP), Volume II: Proliferation Resistance*, DOE/NE-0001/2 (Washington, D.C.: U.S. Government Printing Office, June 1980).

4. "International Nuclear Fuel Cycle Evaluation (INFCE) Report," International Atomic Energy Agency, Vienna, Austria, January 1980.

Chapter 10: The Limits of Technical Fixes

1. A metric ton.

2. Mohamed ElBaradei, statement to the Forty-Fourth Regular Session of the International Atomic Energy Agency General Conference, Vienna, Austria, September 2000.

3. White House Fact Sheet, "Nonproliferation and Export Control Policy Statement," September 27, 1993.

4. U.S. Department of Energy, "A Roadmap for Developing Accelerator Transmutation of Waste Technology," DOE/RW-0519, Washington, D.C., October 1999, p. 4.2.

5. Edwin S. Lyman and Harold A. Feiveson, "The Proliferation Risks of Plutonium Mines," *Science and Global Security* 7 (1998): 119.

6. Feiveson has done a thorough analysis of these different categories and the technologies intended to address each one (Harold A. Feiveson, "Diversion-Resistance Criteria for Future Nuclear Power," Does Nuclear Power Have a Role in Climate Mitigation? workshop, Stanford University, Stanford, Calif., June 22–23, 2000).

7. See, for example, Steve Herring and Philip MacDonald, Idaho National Engineering and Environmental Laboratory, "Low-Cost, Proliferation-Resistant, Uranium-Thorium Dioxide Fuels for Light-Water Reactors," presented at the Workshop on Proliferation-Resistant Nuclear Power Systems, Lawrence Livermore National Laboratory, Livermore, Calif., June 2–4, 1999.

8. "Summary of the Workshop on Proliferation-Resistant Nuclear Power Systems," UCRL-JC-137954, Center for Global Security Research, Lawrence Livermore National Laboratory, Livermore, Calif., June 2–4, 1999, p. 14.

9. Ibid.

10. Ibid.

11. Edwin S. Lyman, "Interim Storage Matrices for Excess Plutonium: Approaching the 'Spent Fuel Standard' without the Use of Reactors," PU/CEES Report No. 286, Center for Energy and Environmental Studies, Princeton University, Princeton, N.J., August 1994.

12. "Technological Opportunities to Increase the Proliferation Resistance of Global Civilian Nuclear Power Systems (TOPS)," report by the TOPS Task

Force of the Nuclear Energy Research Advisory Committee (NERAC), U.S. Department of Energy, Washington, D.C., January 2001. Available at http://www.nuclear.gov/nerac/FinalTOPSRpt.pdf.

13. Amir Rusli and Bakli Arbie, National Atomic Energy Agency for Indonesia, "Identification of Domestic Needs of Modular Heater for Electric and Heat Process Industry in Indonesia," First Information Exchange Meeting on Survey on Basic Studies in the Field of High Temperature Engineering, Nuclear Energy Agency, Paris, September 27–29, 1999.

14. For a comprehensive critique of various transmutation schemes, see Hisham Zerriffi and Arjun Makhijani, "The Nuclear Alchemy Gamble," Institute for Energy and Environmental Research, Takoma Park, Md., August 25, 2000.

15. U.S. Department of Energy, "A Roadmap. . . ."

16. Richard Wagner, Edward Arthur, and Paul Cunningham, "Plutonium, Nuclear Power and Nuclear Weapons," *Perspectives on Science and Technology* (summer 1999).

17. U.S. Department of Energy, "A Roadmap . . . ," p. 4.2.

Chapter 12: Nuclear Power and Nuclear Weapons in India, Pakistan, and Iran

1. Perkovich is the author of the award-winning book, *India's Nuclear Bomb* (Berkeley: University of California Press, 1999), widely considered to be the definitive book on the subject.

2. Homi Bhabha, September 27, 1956, statement at the Conference on the International Atomic Energy Agency Statute, in *Nuclear India*, by J. P. Jain, vol. 2 (New Delhi: Radiant, 1974), pp. 45–46.

Chapter 14: Closing Thoughts on Nonproliferation: The Need for Rigor

1. Hearings before the Senate Committee on Government Operations on S. 1439, 94th Congress, 2nd Session, January 20, 1976, pp. 141 and 142.

2. Ibid., pp. 140 and 143.

3. Ibid., p. 142.

4. Ibid., p. 11.

5. Ibid.

6. Eldon V. C. Greenberg, "The NPT and Plutonium: Application of NPT Prohibitions to 'Civilian' Nuclear Equipment, Technology and Materials Associated with Reprocessing and Plutonium Use," Nuclear Control Institute, Washington, D.C., May 1993.

7. Harold D. Bengelsdorf and Fred McGoldrick, "International Custody of Excess Plutonium," *Bulletin of the Atomic Scientists* (March/April 2002): forthcoming.

8. Luis W. Alvarez, *Adventures of a Physicist* (New York: Basic Books, 1987), p. 125.

9. Paul L. Leventhal, "What Should Be the Fundamental Basis of a National Plutonium Policy?" a paper presented to the International Policy Forum on Management and Disposition of Nuclear Weapons Materials, Leesburg, Va., March 8, 1994.

Chapter 16: An Arms Controller's View

1. The author helped draft the safeguards article when he was in the science bureau of the U.S. Arms Control and Disarmament Agency.

Chapter 17: A Historian's View

The opinions and comments in this chapter are the author's own and do not in any way reflect the policy of the U.S. General Accounting Office, where he works. Previously a journalist, Lanouette has been looking at nuclear issues—both nuclear power and nuclear weapons—since 1969 and covered nuclear proliferation for Dow Jones, *National Journal*, *Atlantic Monthly*, and *Bulletin of the Atomic Scientists*.

1. William Lanouette, *Genius in the Shadows: A Biography of Leo Szilard, The Man Behind the Bomb* (Chicago: University of Chicago Press, 1994).

2. The Pugwash movement, which began in 1957, focuses in part on the prevention of nuclear war, disarmament, an end to nuclear weapons testing, and the nonproliferation of nuclear weapons.

Appendix 1

Originally published in *Preventing Nuclear Terrorism*, edited by Paul Leventhal and Yonah Alexander. Lexington, Mass.: Lexington Books, 1987.

Appendix 2

Originally prepared for a symposium, "The Scope of a Fissile Material Convention," United Nations Institute for Disarmament Research and the Oxford

Research Group, Geneva, Switzerland, August 29, 1996. Revised January 16, 2002.

Alan J. Kuperman is a senior policy analyst for the Nuclear Control Institute and assistant professor of international relations at Johns Hopkins University's School of Advanced International Studies, Bologna, Italy.

1. Luis Alvarez, *Adventures of a Physicist* (Basic Books, 1987), p. 125.

2. Lower-power reactors are less of a concern because they contain less HEU fuel and do not require any fresh HEU fuel during their lifetime. However, in the United States, as an extra precaution, even lower-power licensed research reactors were required to convert to LEU fuel under a 1986 order.

3. The first exception is South Africa's Safari I reactor. However, its operator has now completed a study indicating that conversion is feasible, and the facility's governing board is due to consider conversion in early 2002. The second exception is France's Orphee reactor. However, it is expected to be shut down soon after France commences operation of its new Jules Horowitz reactor, which is designed to use LEU fuel.

4. The U.S. Energy Policy Act of 1992 prohibited export of HEU unless three conditions were satisfied, including such a conversion pledge. The first reactor is the HFR-Petten, in the Netherlands, which will convert to already qualified LEU fuel by 2006. See Ann MacLachlan, "Petten Director Says Study Clears Way for Start of LEU Conversion this Summer," *Nuclear Fuel*, May 31, 1999, p. 6. The second reactor is the BR2, in Belgium. See Ann MacLachlan, "U.S. Agrees to Continue HEU Shipments to BR2 After Belgians Agree to Convert," *Nuclear Fuel*, November 29, 1999. The third reactor is the ILL-Grenoble, in France. See Ann MacLachlan, "U.S. May Resume HEU Fuel Supply as France's ILL Studies Conversion," *Nuclear Fuel*, November 30, 1998, p. 3.

5. In addition, the University of Michigan Ford Nuclear Reactor converted to LEU fuel in 1984, prior to the NRC order. Jim Matos, "U.S. University Reactors Using or Formerly Using HEU Fuel: LEU Conversion Status as of December 2001," RERTR Program, Argonne National Laboratory, undated document obtained January 2002. "Non-Power Reactor HEU to LEU Conversion Program," chart, U.S. Nuclear Regulatory Commission, updated September 29, 1998, by Theodore Michaels, U.S. NRC. The three reactors that cannot yet convert to existing LEU fuels are the university reactors at MIT and University of Missouri–Columbia, and the Department of Commerce's NIST (formerly NBSR) reactor. The small private reactor is operated by General Electric. In the past, some foreign critics have complained about the pace of U.S. conversions. In reality, U.S. conversion efforts have surpassed those overseas. All U.S. research reactors that required fresh HEU fuel, and which can use existing LEU fuel, have been converted or are in the process of doing so, as is true overseas. But in addition, the United States has converted most of its reactors with "lifetime" cores of HEU fuel that do not require refueling, which has not occurred overseas.

The foreign criticism is accurate, however, in that the United States has not credibly demonstrated an intention to convert its remaining five high-power HEU-fueled reactors as soon as suitable LEU fuel is developed. The U.S. Department of Energy could help demonstrate this intention by preparing a conversion feasibility study for its ATR reactor in Idaho.

6. Armando Travelli, RERTR program, Argonne National Laboratory, personal communication to author, January 16, 2002.

7. Shi Yongkang et al., "The China Advanced Research Reactor Project," and Yuan Luzheng et al., "Preliminary Study of Core Characteristics for the Scheduled CARR," presented at the Fifth Meeting of the Asian Symposium on Research Reactors, Taejon, Korea, May 29–31, 1996. A. Ballagny, "The Jules Horowitz Reactor: A new test reactor for fuels and materials," presented at the 1997 International Meeting on Reduced Enrichment for Research and Test Reactors, Jackson Hole, Wyo., October 5–10, 1997.

8. "DOE Facts: A New Neutron Source for the Nation," U.S. Department of Energy, February 1995, p. 1.

9. N. A. Hanan and J. E. Matos, "Fluxes at Experiment Facilities in HEU and LEU Designs for the FRM-II," presented at the 1997 International Meeting on RERTR, Jackson Hole, Wyo., October 5–10, 1997.

10. The director of the FRM-II project, Dr. Klaus Boening, is quoted estimating that a redesign to 32 MW would cost 50–100 million DM. (Jeanne Rubner, "Warnung vor deutschem Sonderweg," *Suddeutsche Zeitung*, April 7, 1998, p. 10.) In light of the fact that he opposes such conversion, this is unlikely an under-estimate. The projected total construction cost of the current design is now at least 720 million DM. Thus, if one accepts Boening's estimate, the extra cost imposed by conversion represents no more than a 7 to 14 percent marginal increase. Such a level of additional cost traditionally has been accepted by states and reactor operators as a necessary and acceptable trade-off for sustaining the nonproliferation and anti-terrorism benefits of LEU fuel and the RERTR program.

11. Letter from Alan J. Kuperman and Paul L. Leventhal to The Honorable Gerhard Schroeder, October 29, 2001, regarding "Risk of Terrorism at Bavaria's FRM-II Reactor." (www.nci.org/01nci/10/schroederletter.htm)

12. Alan J. Kuperman, "A Level Playing Field for Medical Isotope Production—How to Phase Out Reliance on HEU," presented at 22nd International Conference on Reduced Enrichment for Research and Test Reactors, Budapest, Hungary, October 7, 1999. (www.nci.org/q-r/rertr99.htm) See also, Alan J. Kuperman and Paul L. Leventhal, "Forging Consensus to Phase Out HEU for Medical Isotope Production: A Proposed Path Forward," presented at 23rd International Conference on Reduced Enrichment for Research and Test Reactors, Las Vegas, Nev., October 2, 2000. (www.nci.org/q-r/rertr-2000.htm)

Appendix 3

Originally published as "Can Terrorists Build Nuclear Weapons?" in *Preventing Nuclear Terrorism: The Report and Papers of the International Task Force on Prevention of Nuclear Terrorism,* Leventhal and Alexander, eds., Lexington, Mass.: Lexington Books, 1987, pp. 55–65.

J. Carson Mark served as head of the Theoretical Division of Los Alamos National Laboratory and also served on the U.S. Nuclear Regulatory Commission Advisory Committee on Reactor Safeguards and on the Science Advisory Board of the U.S. Air Force.

1. R. Serber, "The Los Alamos Primer," Report L.A. 1, April 1943. (Declassified in 1965.)

2. Quoted by Albert Wohlstetter in *Foreign Policy;* 25, winter, 1976–77; p. 160.

3. Luis W. Alvarez, *Adventures of a Physicist,* New York: Basic Books, 1987, p. 125.

Appendix 4

Marvin M. Miller is a senior research scientist with the Department of Nuclear Engineering and the Center for International Studies at MIT. Miller has served as a Foster fellow with the Nuclear Weapons and Control Bureau of the U.S. Arms Control and Disarmament Agency (ACDA) and is currently a consultant on proliferation issues for ACDA, Los Alamos and Oak Ridge National Laboratories, and Nuclear Control Institute.

1. The development in Sections 11 and 111 follows that of E. V. Weinstock and J. M. de Montmollin, "IAEA Safeguards: Perceptions, Goals and Performance," Workshop on International Safeguards, Cornell University, Ithaca, New York, May 1984.

2. This analogy has been used by other authors. See, in particular, J. Lovett, "International Safeguards for Reprocessing CAN Be Effective," Physics and Society, Vol. 19, No. 3, July 1990, pp. 7–9.

3. To be precise, the error variance of interest is the IAEA inspectors best estimate of MUF, ≈(MUF-D), rather than the variance based on the plant operators measurements alone, ≈(MUF). The difference between these two error variances is not important for the argument here.

4. In practice, unreported plant losses and unmeasured plant inventory could also lead to non-zero values of MUF. Thus, the effect of these error sources must be differentiated from those due to measurement error before the observed non-zero MUF can be unambiguously identified with measurement error.

5. R. Avenhaus and J. Jaech, "On Subdividing Material Balances in Time and/or Space," *Journal of the Institute of Nuclear Materials Management,* Vol. X, No. 3, 1981, pp. 24–33.

BIBLIOGRAPHY

General

Alsema, E. A., P. Frankl, and K. Kato. "Energy Payback Time of Photovoltaic Energy Systems: Present Status and Prospects." In *Proceedings of the Second World Conference and Exhibition on PV Solar Energy Conversion*, edited by J. Schmid et al., Report EUR 188656 EN, Brussels, Belgium, 1998.

Alvarez, Luis W. *Adventures of a Physicist*. New York: Basic Books, 1987.

Bachu, S. "Geological Sequestration of Anthropogenic Carbon Dioxide: Applicability and Current Issues," pp. 285–303. In *Geological Perspectives of Global Climate Change*, edited by L. C. Gerhard, W. E. Harrison, and B. M. Hanson. AAPG Studies in Geology 47, American Association of Petroleum Geologists, Tulsa, Okla., 2001.

Bachu, S., and W. D. Gunter. "Storage Capacity of Carbon Dioxide in Geological Media in Sedimentary Basins with Application to the Alberta Basin," pp. 195–200. In *Greenhouse Gas Control Technologies: Proceedings of the 4th International Conference on GHG Control Technologies*. Amsterdam: Pergamon, 1999.

Bachu, S., W. D. Gunter, and E. H. Perkins. "Aquifer Disposal of CO_2: Hydrodynamic and Mineral Trapping." *Energy Conversion and Management* 35 (1994): 269–79.

Bengelsdorf, Harold D., and Fred McGoldrick. "International Custody of Excess Plutonium." *Bulletin of the Atomic Scientists* (March/April 2002): forthcoming.

Bradley, Robert L. Jr. "Renewable Energy." *Cato Policy Analysis* (280)(August 27, 1997): 21.

British Nuclear Fuels Ltd. Director of Safety, Health, and Environment. "Annual Report on Discharges and Monitoring of the Environment." Risley, Warrington, Cheshire, WA3 6AS, United Kingdom, 1997.

British Nuclear Industrial Forum. "The BNIF Foresight Programme: An Associate Programme of the National Foresight 2000 Programme." London, November 9, 2000.

Brooks, Frederick P. *The Mythical Man-Month*. Reading, Mass.: Addison Wesley Longman, 1982.

Browning, William D., and Joseph J. Romm. "Greening the Building and the Bottom Line." Rocky Mountain Institute, Snowmass, Colo., 1998. Available at www.rmi.org/images/other/GDS-GBBL.pdf.

Byrer, C. W., and H. D. Guthrie. "Coal Deposits: Potential Geological Sink for Sequestering Carbon Dioxide Emissions from Power Plants," pp. 181–187. In *Greenhouse Gas Control Technologies: Proceedings of the 4th International Conference on GHG Control Technologies*. Amsterdam: Pergamon, 1999.

Cabraal, A., M. Cosgrove-Davies, and L. Schaeffer. "Best Practices for Photovoltaic Household Electrification Programs: Lessons from Experiences in Selected Countries." Technical Paper No. 324, World Bank, Washington, D.C., 1996.

Cavallo, A. J. "High-Capacity Factor Wind Energy Systems." *Journal of Solar Energy Engineering* 117 (1995): 137–43.

Chang, Y. I., and C. E. Till. "The Integral Fast Reactor." *Advances in Nuclear Science & Technology* 20 (1988): 127–54.

Charpak, G., and R. L. Garwin. *Feux Follets et Champignons Nucleaires*. Paris: Editions Odille Jacob, 1998.

Chayes, A., and W. B. Lewis. *International Arrangements for Nuclear Fuel Reprocessing*. Cambridge, Mass.: Ballinger Press, 1977.

Clawson, Patrick L., ed. *Energy and National Security in the 21st Century*. Washington, D.C.: National Defense University Press, 1995.

Comby, Bruno. *Environmentalists for Nuclear Energy*. Available at http://www.ecolo.org/base/baseen.htm.

ElBaradei, Mohamed. Statement to the Forty-fourth Regular Session of the International Atomic Energy Agency General Conference, Vienna, Austria, September 2000.

Electric Power Research Institute. "Technical Assessment Guide: Electricity Supply—1993." EPRI TR-102275-V1R7, vol. 1, rev. 1, Palo Alto, Calif., June 1993.

Electric Power Research Institute and U.S. Department of Energy. Energy Efficiency and Renewable Energy. Office of Utility Technologies. "Renewable Energy Technology Characterizations." EPRI TR-10949, Electric Power Research Institute, Palo Alto, Calif., December 1997.

Elliott, D. L., L. L. Wendell, and G. L. Gower. "An Assessment of the Available Windy Land Area and Wind Energy Potential in the Contiguous United States." Pacific Northwest Laboratories Report PNL-7789, Pacific Northwest Laboratory, Richland, Wash., 1991.

Energy R&D Panel of the President's Committee of Advisors on Science and Technology. "Federal Energy Research & Development for the Challenges of the 21st Century." Office of Science and Technology Policy, Executive Office of the President, Washington, D.C., November 1997. Available at http://www.whitehouse.gov/WH/EOP/OSTP/html/ISTP_Home.html.

E SOURCE (Boulder, Colo.). Semiannual *Electronic Encyclopedia*, www.esource.com.

Feiveson, Harold A. "Comments on the Development Path for Proliferation-Resistant Nuclear Power." Paper presented at a conference at the James Baker Institute, Rice University, Houston, Texas, March 2001.

———. "Diversion-Resistance Criteria for Future Nuclear Power." Does Nuclear Power Have a Role in Climate Mitigation? Workshop, Stanford University, Stanford, Calif., June 22–23, 2000.

———, ed. *Nuclear Turning Point: A Blueprint for Deep Cuts and De-Alerting of Nuclear Weapons*. Washington, D.C.: Brookings Institution, 1999.

Filin, A. I., V. V. Orlov, V. N. Leonov, A. G. Sila-Novitskij, V. S. Smirnov, and V. S. Tsikunov. "Design Features of BREST Reactors; Experimental Work to Advance the Concept of BREST Reactors; Results and Plans." Paper presented at the Global '99: Nuclear Technology—Bridging the Millennia International Conference on Future Nuclear Systems, Jackson Hole, Wyo., August 29–September 3, 1999.

Foos, Jacques et al. "Document of 1996 on Extraction of Uranium from Seawater." COGEMA, June 1996.

Galperin, A., P. Reichert, and A. Radkowsky. "Thorium Fuel for Light Water Reactors—Reducing the Proliferation Potential of the Nuclear Fuel Cycle." *Science & Global Security* 6 (1997): 265–90.

Garwin, Richard L. "Post–Cold War World and Nuclear Weapons Proliferation." Paper delivered at Session 5, "Nuclear Nonproliferation and Plutonium," 29th Japan Atomic Industrial Forum (JAIF) Annual Conference, Nagoya, Japan, April 19, 1996.

———. "Reactor-Grade Plutonium Can Be Used to Make Powerful and Reliable Nuclear Weapons: Separated Plutonium in the Fuel Cycle Must Be Protected as If It Were Nuclear Weapons." Newsletter of the Nuclear Information Service, Japan, August 26, 1998. Available at www.fas.org/rlg.

———. "L'uranium extrait de l'eau de mer: un combustible vert pour demain? [Seawater uranium: a green fuel for the future?]." In "Les Energies du Futur" [Energies of the Future]. *Revue des Deux Mondes* (Paris) (April 2001): 67–77.

———. Letter in *Physics Today* 53 (5)(May 2000): 12–14.

Gerhard, L. C., W. E. Harrison, and B. M. Hanson, eds. *Geological Perspectives of Global Climate Change*. AAPG Studies in Geology 47, American Association of Petroleum Geologists, Tulsa, Okla., 2001.

Goodwin, B., and J. Kammerdiener. "Future Proliferation Threat." Paper presented the Proliferation-Resistant Nuclear Power Systems Workshop, Center for Global Security Research, Lawrence Livermore National Laboratory, Livermore, Calif., June 2–4, 1999. Available at http://cgsr.llnl.gov/Final_Wkshp_Rpt.pdf.

Greenhouse Gas Control Technologies: Proceedings of the 4th International Conference on GHG Control Technologies, edited by B. Eliasson, P. Riemer, and A. Wokaun. Amsterdam: Pergamon, 1999.

Grübler, Arnulf. *Technology and Global Change*. Cambridge: Cambridge University Press, 1998.

Grübler, Arnulf, Nebojsa Nakicenovic, and David G. Victor. "Dynamics of Energy Technologies and Global Change." *Energy Policy* 27 (1999): 265.

Gunter, W. D., R. J. Chalaturnyk, and J. D. Scott. "Monitoring of Aquifer Disposal of Carbon Dioxide: Experience from Underground Gas Storage and Enhanced Oil Recovery," pp. 151–56. In *Greenhouse Gas Control Technologies: Proceedings of the 4th International Conference on GHG Control Technologies.* Amsterdam: Pergamon, 1999.

Gunter, W. D., E. H. Perkins, and T. J. McCann. "Aquifer Disposal of Carbon Dioxide–Rich Gases: Reaction Design for Added Capacity." *Energy Conversion and Management* 34 (1993): 941–48.

Gunter, W. D., T. Gentzix, B. A. Rottenfusser, and R. J. H. Richardson. "Deep Coalbed Methane in Alberta, Canada: A Fuel Resource with the Potential of Zero Greenhouse Emissions." *Energy Conversion and Management* 38 (1997): S217–22.

Hawken, Paul, Amory B. Lovins, and L. Hunter Lovins. *Natural Capitalism: Creating the Next Industrial Revolution.* New York: Little, Brown, 1999.

Hendriks, C. A. "Carbon Dioxide Removal from Coal-Fired Power Plants." Ph.D. thesis, Department of Science, Technology, and Society, Utrecht University, Utrecht, Netherlands, 1994.

Herring, Steve, and Philip MacDonald. Idaho National Engineering and Environmental Laboratory. "Low-Cost, Proliferation-Resistant, Uranium-Thorium Dioxide Fuels for Light-Water Reactors." Paper presented at the Proliferation-Resistant Nuclear Power Systems Workshop, Center for Global Security Research, Lawrence Livermore National Laboratory, Livermore, Calif., June 2–4, 1999. Available at http://cgsr.llnl.gov/Final_Wkshp_Rpt.pdf.

Hill, R. N., J. E. Cahalan, H. S. Khalil, and D. C. Wade. "Development of Small, Fast Reactor Core Designs Using Lead-based Coolant." Paper presented at the Global '99: Nuclear Technology—Bridging the Millennia International Conference on Future Nuclear Systems, Jackson Hole, Wyo., August 29–September 3, 1999.

Hitchon, B., W. D. Gunter, T. Gentzis, and R. Bailey. "Sedimentary Basins and Greenhouse Gases: A Serendipitous Association." *Energy Conversion and Management* 40 (1999): 825–43.

Hoffert, M. I., K. Caldeira, A. K. Jain, E. F. Haites, L. D. D. Harvey, S. D. Potter, M. E. Schlessinger, S. H. Schneider, R. G. Watts, T. M. L. Wigley, and D. J. Wuebbles. "Energy Implications of Future Stabilization of Atmospheric CO_2 Content." *Nature* 329 (October 29, 1998): 881–84.

Holdren, J. P., and K. R. Smith, "Energy, the Environment, and Health," chapter 3. In World Energy Assessment. *Energy and the Challenge of Sustainability.* New York: Bureau for Development Policy, UN Development Programme, 2000.

Holloway, S. "Safety of Underground Disposal of Carbon Dioxide." *Energy Conversion and Management* 38 (1997): S241–45.

———, ed. "The Underground Storage of Carbon Dioxide." Report prepared for the Joule II Programme (DG XII) of the Commission of the European Communities, Contract No. JOU2 CT92–0031, Brussels, Belgium, February 1996.

Intergovernmental Panel on Climate Change. *Aviation and the Global Atmosphere.* Oxford: Oxford University Press, 1999.

———. *Climate Change 1994—Radiative Forcing of Climate Change and an Evaluation of the IPCC IS92 Scenarios.* Cambridge: Cambridge University Press, 1994.

———. *Climate Change 1995: Impacts, Adaptations and Mitigation of Climate Change: Scientific-Technical Analyses.* Second assessment report of the Intergovernmental Panel on Climate Change. Cambridge: Cambridge University Press, 1996.

International Atomic Energy Agency. "International Nuclear Fuel Cycle Evaluation (INFCE) Report." Vienna, Austria, 1978.

International Council for Science. *An Agenda of Science for Environment and Development into the 21st Century.* Cambridge: Cambridge University Press, 1992.

International Energy Agency. "Carbon Dioxide Capture and Storage in the Natuna NG [Natural Gas] Project." *Greenhouse Issues* 22 (1996): 1.

Ippolito, T. D. Jr. "Effects of Variation of Uranium Enrichment on Nuclear Submarine Reactor Design." Master's degree thesis, Department of Nuclear Engineering, Massachusetts Institute of Technology, Cambridge, Mass., May 1990.

Jain, J. P. *Nuclear India.* Vol. 2. New Delhi: Radiant, 1974.

Johansson, Thomas B., Henry Kelly, Amulya K. N. Reddy, and Robert H. Williams, eds. *Renewable Energy: Sources for Fuels and Electricity.* Washington, D.C.: Island Press, 1993.

Kaarstad, O. "Emission-Free Fossil Energy from Norway." *Energy Conversion and Management* 33 (5–8)(1992): 781–86.

Kato, T. et al. "Conceptual Design of Uranium Recovery Plant from Seawater." *Journal of Thermal and Nuclear Power Engineering Society* 50 (1999): 71–77. In Japanese.

Kato, T., K. Okugawa, Y. Sugihara, and T. Matsumura, "Conceptual Design of Uranium Recovery Plant from Seawater," undated (perhaps March 2001).

Keeny, Spurgeon M. Jr. et al. *Nuclear Power Issues and Choices.* Cambridge, Mass.: Ballinger Press, March 1977.

Lanouette, William. *Genius in the Shadows: A Biography of Leo Szilard, The Man Behind the Bomb.* Chicago: University of Chicago Press, 1994.

Lew, D., R. H. Williams, S. Xie, and S. Zhang. "Large-scale Baseload Wind Power in China." *Natural Resources Forum* 22 (3)(1998): 165–218.

Lewis, J. S., and R. W. Niedzwiecki. "Aircraft Technology and Its Relation to Emissions," chapter 7. In Intergovernmental Panel on Climate Change. *Aviation and the Global Atmosphere.* Oxford: Oxford University Press, 1999.

Longworth, H. L., G. C. Dunn, and M. Semchuck. "Underground Disposal of Acid Gas in Alberta, Canada: Regulatory Concerns and Case Histories." SPE 35584, paper presented at the Gas Technology Conference, Calgary, Alberta, Canada, April 28–May 1, 1996.

Lopatkin, A. V., and V. V. Orlov. "Fuel Cycle of BREST 1200 with Non-Proliferation of Plutonium and Equivalent Disposal of Radioactive Waste." Paper presented at the Global '99: Nuclear Technology—Bridging the Millennia International Conference on Future Nuclear Systems, Jackson Hole, Wyo., August 29–September 3, 1999.

Lovelock, James. "Preface." In *Environmentalists for Nuclear Energy*, by Bruno Comby. Paris: TNR Editions. Available at http://www.ecolo.org/lovelock/loveprefaceen.htm.

Lovins, Amory B. "Energy Strategy: The Road Not Taken?" *Foreign Affairs* (October 1976).

———. "Profiting from a Nuclear-Free Third Millennium." *Power Economics* (November 1999).

———. *Soft Energy Paths: Toward a Durable Peace*. Cambridge, Mass.: Ballinger Press, 1977.

———. "The Super-Efficient Passive Building Frontier." *ASHRAE Journal* (June 1995).

Lovins, Amory B., and L. Hunter Lovins. "Climate: Making Sense *and* Making Money." Rocky Mountain Institute, Snomass, Colo., 1997–1998. Available at www.rmi.org/images/other/C-ClimateMSMM.pdf.

———. "Fool's Gold in Alaska." *Foreign Affairs* (July/August 2001). Annotated at www.rmi.org/images/other/E-FoolsGoldAnnotated.pdf.

Lovins, Amory B., and John H. Price. *Non-Nuclear Futures: The Case for an Ethical Energy Strategy*. Cambridge, Mass.: Ballinger Press, 1975.

Lovins, Amory B., and Brett D. Williams. "A Strategy for the Hydrogen Transition." National Hydrogen Association, Vienna, Va., April 1999. Available at www.rmi.org/images/other/HC-StrategyHCTrans.pdf.

Lovins, Amory B., L. Hunter Lovins, and Leonard Ross. "Nuclear Power and Nuclear Bombs." *Foreign Affairs* (summer 1980).

Lyman, Edwin S. "Interim Storage Matrices for Excess Plutonium: Approaching the 'Spent Fuel Standard' without the Use of Reactors." PU/CEES Report No. 286, Center for Energy and Environmental Studies, Princeton University, Princeton, N.J., August 1994.

Lyman, Edwin S., and Harold A. Feiveson. "The Proliferation Risks of Plutonium Mines." *Science and Global Security* 7 (1998): 119.

MacLachlan, Anne. "Eurodif Amortization 'Challenges' Cost of MOX Fuel for EDF." *NuclearFuel*, January 22, 2001, p. 4.

Mark, J. Carson, Theodore Taylor, Eugene Eyster, William Maraman, and Jacob Wechsler. "Can Terrorists Build Nuclear Weapons?" pp. 55–65. In *Preventing Nuclear Terrorism*, edited by Paul Leventhal and Yonah Alexander. Lexington, Mass.: Lexington Books, 1987.

Markey, Edward J., and Douglas C. Waller. *Nuclear Peril: The Politics of Proliferation.* New York: HarperInformation, January 1982.

Marnay, C., R. C. Richey, S. A. Mahler, and R. J. Markel. "Estimating the Environmental and Economic Effects of Widespread Residential PV Adoption Using GIS and NEMS." Paper presented at the 1997 American Solar Energy Society Meeting, Washington, D.C., May 1997.

Modis, Theodore. *Predictions.* New York: Simon & Schuster, 1992.

Morehart, M., J. Ryan, D. Peacock, and R. Strickland. "U.S. Farm Income Decline in 2000 to Be Tempered by Government Payments." *Agricultural Outlook* (Economic Research Service, U.S. Department of Agriculture) (January–February, 2000).

Nadel, Steve. "Lessons Learned." Rep. 90–8, New York State Energy Research and Development Authority, New York State Energy Office, Niagara Mohawk Power Corp., and American Council for an Energy-Efficient Economy, Albany, N.Y., 1990.

Nobukawa, H. et al. "Development of a Floating Type System for Uranium Extraction from Seawater Using Sea Current and Wave Power." *Proceedings of the 4th International Offshore and Polar Engineering Conference*, Osaka, Japan, April 10–15, 1994, pp. 294–300.

Ogden, J. M., and R. H. Williams. *Solar Hydrogen: Moving beyond Fossil Fuels.* Washington, D.C.: World Resources Institute, 1989.

Ogden, J. M., and K. Yoshida. "Present Status of R&D for Hydrogen Production from Water in Japan." *Energy Research* 7 (1983): 1–12.

Ogden, J. M., R. H. Williams, and E. D. Larson. "Toward a Hydrogen-Based Transportation System." Draft manuscript, Princeton Environmental Institute, Princeton University, Princeton, N.J., May 2001.

Organisation for Economic Co-operation and Development. Nuclear Energy Agency. "Reductions of Capital Costs of Nuclear Power Plants." Paris, 2000.

———. "Technical Appraisal of the Current Situation in the Field of Waste Management: A Collective Opinion by the Reactor Waste Management Committee." Paris, 1985.

Orlov, V., V. Leonov, A. Sila-Novitski, V. Smirnov, V. Tsikunov, and A. Filin. "Nuclear Power of the Coming Century and Requirements of the Nuclear Technology." Paper presented at the Global '99: Nuclear Technology—Bridging the Millennia International Conference on Future Nuclear Systems, Jackson Hole, Wyo., August 29–September 3, 1999.

Ormerod, W. "The Disposal of Carbon Dioxide from Fossil Fuel Power Stations." IEA/GHG/SR3, IEA Greenhouse Gas Research and Development Programme, Cheltenham, United Kingdom, 1994.

Paffenbarger, J. A., and E. Bertel. "Results from the OECD Report on International Projects of Electricity Generating Costs." Paper presented at IJPGC 98: International Joint Power Generation Conference and Exhibition, Baltimore, Md., August 24–26, 1998.

Panofsky, Wolfgang K. H.. "Management and Disposition of Excess Weapons Plutonium." Report of the Committee on International Security and Arms Control, National Academy of Sciences, Washington, D.C., January 1994, pp. 32–33.

Pasternak, Alan D. "Global Energy Futures and Human Development: A Framework for Analysis." UCRL-ID-140773, Lawrence Livermore National Laboratory, Livermore, Calif., 2001.

Payne, A., R. Duke, and Robert H. Williams. "Accelerating Residential PV Expansion: Supply Analysis for Competitive Electricity Markets." *Energy Policy* 29 (2001): 787–800.

PCAST Panel on International Cooperation in Energy Research, Development, Demonstration, and Deployment. *Powerful Partnerships: The Federal Energy Research & Development for the Challenges of the 21st Century*. Report of the Panel on International Cooperation in Energy Research, Development, Demonstration, and Deployment of the President's Committee of Advisors on Science and Technology, Washington, D.C., June 1999. Available at http://www.whitehouse.gov/WH/EOP/OSTP/html/ISTP_Home.html.

Peiyan, Zeng. Director, State Development Planning Commission. Press statement dated March 6, 2000, as reported in *Zhongguo Dianli Bao* (China Electric Power Daily), March 9, 2000.

Perkovich, George. *India's Nuclear Bomb*. Berkeley: University of California Press, 1999.

Report to the American Physical Society by the Study Group on Nuclear Fuel Cycles and Waste Management." *Reviews of Modern Physics* 50 (1, Part 2)(January 1978): S156.

Rhodes, Richard. *Dark Sun*. New York: Simon & Schuster, 1995.

———. *The Making of the Atomic Bomb*. New York: Simon & Schuster, 1986.

Rhodes, Richard, and Denis Beller. "The Need for Nuclear Power." *Foreign Affairs* (January/February 2000).

Rogner, H.-H. "Energy Resources," chapter 5. In World Energy Assessment. *Energy and the Challenge of Sustainability*. New York: Bureau for Development Policy, UN Development Programme, 2000.

Rusli, Amir, and Bakli Arbie. National Atomic Energy Agency for Indonesia. "Identification of Domestic Needs of Modular Heater for Electric and Heat Process Industry in Indonesia." First Information Exchange Meeting on Survey on Basic Studies in the Field of High Temperature Engineering, Nuclear Energy Agency, Organisation for Economic Co-operation and Development, Paris, September 27–29, 1999.

Schmid, J. et al., eds. *Proceedings of the Second World Conference and Exhibition on PV Solar Energy Conversion*. Report EUR 188656 EN. Brussels, Belgium, 1998.

Shainker, Robert B. Electric Power Research Institute. Presentation to the PCAST Panel on International Cooperation in Energy Research, Development, Demonstration, and Deployment, July 14, 1997. Reproduced from Panel on International Cooperation in Energy Research, Development, Demonstration, and Deployment. President's Committee of Advisors on Science and Technology. *Powerful Partnerships: The Federal Energy Research & Development for the Challenges of the 21st Century.* Washington, D.C.: Office of Science and Technology Policy, Executive Office of the President, June 1999. Available at http://www.whitehouse.gov/WH/EOP/OSTP/html/ISTP/Home.html.

Shainker, Robert B., B. Mehta, and R. Pollack. "Overview of CAES Technology," pp. 992–97. In *Proceedings of the American Power Conference.* Chicago: Illinois Institute of Technology, 1993.

Shapiro, R. J. *Statistical Abstract of the United States.* 120th ed. Washington, D.C.: Bureau of the Census, U.S. Department of Commerce.

Sims, Gordon. *The Anti-Nuclear Game.* Ottawa: University of Ottawa Press, 1990.

Socolow, R. H., ed. "Fuels Decarbonization and Carbon Sequestration: Report of a Workshop by the Members of the Report Committee." PU/CEES Report 302, Center for Energy and Environmental Studies, Princeton University, Princeton, N.J., September 1997. Available at http://www.princeton.edu/~ceesdoe.

Steinfeld, A., and R. Palumbo. "Fuels from Sunlight and Water." Paul Scherrer Institute, Switzerland, 2001. Available at www.psi.ch.

Stevens, S. H., V. A. Kuuskraa, and J. Gale. "Sequestration of Carbon Dioxide in Depleted Oil and Gas Fields: Global Capacity, Costs, and Barriers," pp. 278–83. In *Greenhouse Gas Control Technologies: Proceedings of the 5th International Conference on GHG [Greenhouse Gas] Control Technologies (GHGT-5),* edited by D. J. Williams, R. A. Durie, P. McMullan, C. A. J. Paulson, and A. Y. Smith. Collingwood, Victoria, Australia: CSIRO Publishing, 2000.

Stevens, S. H., V. A. Kuuskraa, D. Spector, and P. Riemer. "Enhanced Coalbed Methane Recovery Using Carbon Dioxide Injection: Worldwide Resource and Carbon Dioxide Injection Potential," pp. 175–180. *Greenhouse Gas Control Technologies: Proceedings of the 4th International Conference on GHG Control Technologies,* edited by B. Eliasson, P. Riemer, and A. Wokaun. Amsterdam: Pergamon, 1999.

Summerfield, I. R., S. H. Goldthorpe, N. Williams, and A. Sheikh. "Costs of Carbon Dioxide Disposal Options." In *Proceedings of the International Energy Agency Carbon Dioxide Disposal Symposium.* Amsterdam: Pergamon, 1993.

Tadokoro, Y., T. Kajiyama, T. Yamaguchi, N. Sakai, H. Kameyama, and K. Yoshida. "Technical Evaluation of UT-3 Thermochemical Hydrogen Production Process for an Industrial Scale Plant." *International Journal of Hydrogen Energy* 22 (1)(1997): 49–56.

Taylor, J. T. "Economic and Market Potential of Small Innovative Reactors." Paper presented at the New Energy Technologies: A Policy Framework for Micro-Nuclear Technology Workshop, Houston, Texas, March 19–20, 2001.

Teller, Edward. *Energy From Heaven and Earth.* San Francisco: W. H. Freeman, 1979.

———. "Fast Reactors: Maybe." *Nuclear News*, August 21, 1967.

Tuchman, Barbara W. *The March of Folly*. New York: Alfred A. Knopf, 1984.

United Nations. Special Committee on the Effects of Atomic Radiation. "Sources and Effects of Ionizing Radiation." Report to the General Assembly, with Scientific Annexes. New York: United Nations, 1993.

United States. Central Intelligence Agency. *World Factbook.* Washington, D.C.: Central Intelligence Agency, July 1, 2001.

United States. Congress. Hearings before the Senate Committee on Government Operations on S. 1439, 94th Congress, 2nd Session, Washington, D.C., January 20, 1976.

United States. Department of Energy. "Final Nonproliferation and Arms Control Assessment of Weapons-Usable Fissile Material Storage and Excess Plutonium Disposition Alternatives." Washington, D.C., 1997.

———. "Nonproliferation and Arms Control Assessment of Weapons-Usable Fissile Material Storage and Disposition Alternatives." Draft, Washington, D.C., October 1996.

———. *Nuclear Proliferation and Civilian Nuclear Power: Report of the Nonproliferation Alternative Systems Assessment Program (NASAP), Volume II: Proliferation Resistance.* DOE/NE-0001/2. Washington, D.C.: U.S. Department of Energy, 1980.

———. "A Roadmap for Developing Accelerator Transmutation of Waste (ATW) Technology: A Report to Congress." DOE/RW-0519, Washington, D.C., October 1999.

United States. Department of Energy. Energy Information Administration. "Annual Energy Outlook 2001: With Projections to 2020." DOE/EIA-0383 (2001). Washington, D.C., December 2000.

———. "International Energy Outlook 2001." DOE/EIA-0484 (2001), Washington, D.C., March 2001.

United States. Department of Energy. Nuclear Energy Research Advisory Committee. Technical Opportunities for Increasing the Proliferation Resistance of Nuclear Power Systems Task Force. "Annex: Attributes of Proliferation Resistance for Civilian Nuclear Power Systems." Washington, D.C., October 2000. Available at http://www.nuclear.gov/nerac/FinalTOPSRptAnnex.pdf.

United States. Department of Energy. Nuclear Energy Research Advisory Committee. Technical Opportunities for Increasing the Proliferation Resistance of Nuclear Power Systems Task Force. "Report of the International Workshop on Technology Opportunities for Increasing the Proliferation Resistance of Global Civilian Nuclear Power Systems, March 29–30, 2000." Washington, D.C. Available at www.nuclear.gov.

United States. Department of Energy. Nuclear Energy Research Advisory Committee. Technical Opportunities for Increasing the Proliferation Resistance of

Nuclear Power Systems Task Force. "Technological Opportunities to Increase the Proliferation Resistance of Global Civilian Nuclear Power Systems." Washington, D.C., January 2001. Available at http://www.nuclear.gov/nerac/FinalTOPSRpt.pdf.

United States. Department of State. Board of Consultants to the Secretary of State's Committee on Atomic Energy. *A Report on the International Control of Atomic Energy*. U.S. Department of State Publication 2498. Washington, D.C.: U.S. Government Printing Office, 1946.

United States. White House. "Nonproliferation and Export Control Policy Statement." Fact Sheet, Washington, D.C., September 27, 1993.

van der Burgt, M. J., J. Cantle, and V. K. Boutkan. "Carbon Dioxide Disposal from Coal-Based IGCCs [Integrated Gasifier/Combined Cycles] in Depleted Gas Fields." *Energy Conversion and Management* 33 (5–8)(1992): 603–10.

Wagner, Richard L. Jr., Edward Arthur, and Paul Cunningham. "Future Nuclear Energy: Ensuring a U.S. Place at the International Table." Unpublished draft, Los Alamos National Laboratory, Los Alamos, N.M., August 4, 1997.

———. "Plutonium, Nuclear Power and Nuclear Weapons." *Perspectives on Science and Technology* (summer 1999).

Weinberg, A. W. "From Technological Fixer to Think-Tanker." *Annual Review of Energy and the Environment*. Annual Reviews Palo Alto, Calif., 1994.

Wichert, E., and T. Royan. "Acid Gas Injection Eliminates Sulfur Recovery Expense." *Oil and Gas Journal* (April 28,1997): 67–72.

Wigley, T. M. L., R. Richels, and J. A. Edmonds. "Economic and Environmental Choices in the Stabilization of Atmospheric CO_2 Concentration." *Nature* 379 (January 18, 1996): 240–43.

Williams, Robert H. "Toward Zero Emissions for Transportation Using Fossil Fuels." Paper prepared for Managing Transitions in the Transportation Sector: How Fast and How Far? VIII Biennial Conference on Transportation, Energy, and Environmental Policy, Asilomar Conference Center, Monterey, Calif., September 11–14, 2001.

———. "Toward Zero Emissions for Coal: Roles for Inorganic Membranes," pp. 212–42. In *Proceedings of the International Symposium toward Zero Emissions: The Challenge for Hydrocarbons*. Rome: EniTecnologie, March 1999.

Williams, Robert H., and G. Terzian. "A Benefit/Cost Analysis of Photovoltaic Technology." PU/CEES Report No. 281, Center for Energy and Environmental Studies, Princeton University, Princeton, N.J., October 1993.

World Energy Assessment. *Energy and the Challenge of Sustainability*. New York: Bureau for Development Policy, UN Development Programme, 2000.

World Energy Council. *New Renewable Energy Resources: A Guide to the Future*. London: Kogan Page, 1994.

Worrell, E. "Advanced Technologies and Energy Efficiency in the Iron and Steel Industry in China." *Energy for Sustainable Development* II (4)(November 1995): 27–40.

Yalçin, S. "A Review of Nuclear Hydrogen Production." *International Journal of Hydrogen Energy* 14 (8)(1989): 551–61.

Yoshida, K. "Present Status of R&D for Hydrogen Production from Water in Japan." *Energy Research* 7 (1983): 1–12.

Zerriffi, Hisham, and Arjun Makhijani. "The Nuclear Alchemy Gamble." Institute for Energy and Environmental Research, Takoma Park, Md., August 25, 2000.

Zrodnikov, A. V., V. I. Chitaykin, B. F. Gromov, G. I. Toshinsky, U. G. Dragunov, and V. S. Stepanov. "Application of Reactors Cooled by Lead-Bismuth Alloy in Nuclear Power Energy." Paper presented at the Global '99: Nuclear Technology—Bridging the Millennia International Conference on Future Nuclear Systems, Jackson Hole, Wyo., August 29–September 3, 1999.

Nuclear Control Institute

Albright, David H. "World Inventories of Civilian Plutonium and the Spread of Nuclear Weapons." Nuclear Control Institute Special Report, Nuclear Control Institute, Washington, D.C., 1983.

"Asia-Pacific Forum on Sea Shipments of Japanese Plutonium: Issues and Concerns." Nuclear Control Institute and Citizens' Nuclear Information Service, Tokyo, Japan, October 1992.

Bethe, Hans, Freeman Dyson, Herman Feshbach, Val Fitch, Marvin Goldberger, Kurt Gottfried, Milton Hoenig, Franklin Long, J. Carson Mark, George Rathjens, and Victor Weisskopf. "Atomic Weapons: America's Non-Existent Nuclear 'Crisis.'" *Washington Post*, Outposts page, Outlook section, April 16, 1989.

Buell, John. "The Argentine Enrichment Plant." Nuclear Control Institute Issue Brief, Washington, D.C., December 1983.

Dolley, Steven. "Japanese Strategic Uranium Reserve: A Response to BNFL." *Nuclear Engineering International* (September 1994): 50–51.

———. "Iraq and the Bomb: The Nuclear Threat Continues." Nuclear Control Institute, Washington, D.C., February 19, 1998. Available at http://www.nci.org/i/ib21998.htm.

———. "Iraq's Nuclear Weapons Program: Unresolved Issues." Nuclear Control Institute, Washington, D.C., May 12, 1998. Available at http://www.nci.org/iraq/iraq511.htm.

Gilinsky, Victor, and Paul L. Leventhal. "India Cheated." *Washington Post*, June 15, 1998, op-ed page. Available at http://www.nci.org/a/a61598-txt.htm.

Greenberg, Eldon V. C. "The NPT and Plutonium: Application of NPT Prohibitions to 'Civilian' Nuclear Equipment, Technology and Materials Associated with Reprocessing and Plutonium Use." Nuclear Control Institute, Washington, D.C., May 1993. (Update of Memorandum, July 29, 1985.)

———. "Japanese Assurances Concerning Subsequent Retransfers and Utilization of Reprocessed Plutonium." Nuclear Control Institute, Washington, D.C., February 12, 1986.

———. "Legal Deficiencies in the Proposed Agreement for Nuclear Cooperation between the United States and Japan." Nuclear Control Institute, Washington, D.C., January 29, 1988.

———. "Legal Deficiencies in the U.S.–China Agreement for Nuclear Cooperation: A Comment on the Executive Branch's Analysis." Nuclear Control Institute, Washington, D.C., September 20, 1985.

———. "Legal Deficiencies in the U.S.–China Agreement for Nuclear Cooperation and the Need for Enhanced Congressional Review." Nuclear Control Institute, Washington, D.C., September 9, 1985.

———. "Rebuttal to the Executive Branch 'Fact Sheet' Concerning Misconceptions about the U.S.–Japan Nuclear Agreement." Nuclear Control Institute, Washington, D.C., January 1988.

———. "Use of Plutonium Produced by the *Superphenix* Reactor for Military Purposes." Nuclear Control Institute, Washington, D.C., January 4, 1985.

Horner, Daniel. "Curbing Proliferation: An Agenda for the NPT Review Conference." Nuclear Control Institute, Washington, D.C., 1985.

———. "Strengthening Safeguards: A High Priority for the NPT Review Conference." Nuclear Control Institute, Washington, D.C. 1985.

Kuperman, Alan J. "A Level-Playing Field for Medical Isotope Production: How to Phase Out Reliance on Highly Enriched Uranium." Paper presented at 22nd International Meeting on Reduced Enrichment for Research and Test Reactors (RERTR), Budapest, Hungary, October 1999. Available at http://www.nci.org/q-r/rertr99.htm.

———. "Civilian Highly Enriched Uranium and the Fissile Material Convention." Paper prepared for The Scope of a Fissile Material Convention Symposium, UNIDIR and Oxford Research Group, Geneva, Switzerland, August 29, 1996, and updated October 1998. Available at http://www.nci.org/i/ib82996.htm.

Kuperman, Alan J., and Paul L. Leventhal. "RERTR End-Game: A Win-Win Framework." Paper presented at the International Meeting on the Reduced Enrichment for Research and Test Reactors (RERTR) Program, Jackson Hole, Wyo., October 5, 1997. Available at http://www.nci.org/i/ib10597.htm.

———. "Nuclear Proliferation Is Everybody's Business." *Los Angeles Times*, December 25, 1988, op ed page.

———. "Forging Consensus to Phase Out Highly Enriched Uranium for Medical Isotope Production: A Proposed Path Forward." Paper presented at Panel on Converting Medical Isotope ProductionRERTR-2000, 23rd International Conference on Reduced Enrichment for Research and Test Reactors (RERTR), Las Vegas, Nev., October 2, 2000.

Leventhal, Paul L. "Cut Off Aid to Pakistan." *Washington Post*, October 8, 1990, op-ed page.

———. "East Asia's Spent Fuel Dilemma." Presented to the 2000 Carnegie International Non-Proliferation Conference, Washington, D.C., March 16, 2000. Available at http://www.nci.org/p/pl-carnegie2000.htm.

———. "Flaws in the Non-Proliferation Treaty." *Bulletin of the Atomic Scientists* (September 1985): 12–15.

———. "France and the Evolution of Proliferation." Paper presented to the Conference on Energy Strategy: Between the Nuclear Risk and the Greenhouse Effect, Palais du Luxembourg, Paris, France, April 8, 1994.

———. "Getting Serious about Proliferation." *Bulletin of the Atomic Scientists* (March 1984): 7–9.

———. "IAEA Safeguards Shortcomings: A Critique." Nuclear Control Institute, Washington, D.C., September 12, 1994. Available at http://www.nci.-org/p/plsgrds.htm.

———. "The Impact of Japan's Plutonium Program on Global Proliferation and Nuclear Terrorism." Paper presented to the Asia-Pacific Forum on Sea Shipments of Japanese Plutonium: Issues and Concerns, Nuclear Control Institute and Citizens' Nuclear Information Service, Tokyo, Japan, October 1992.

———. "International Plutonium Storage: Managing the Unmanageable." Paper presented to a Symposium on the Future of Foreign Nuclear Materials, Naval Postgraduate School, Monterey, Calif., December 8, 1993.

———. "Is Iraq Evading the Nuclear Police?" *New York Times*, December 28, 1990, op-ed page.

———. "Latent and Blatant Proliferation: Does the NPT Work Against Either?" NCI Special Report. NPT at the Crossroads Series, Nuclear Control Institute, Washington, D.C., August 1990.

———. "Latin America, Plutonium and the NPT." Paper presented to the International Conference on the Latin American View on Extension of the NPT, Cordoba, Argentina, October 21, 1994.

———. "Let's Rethink Plutonium Use." *Asahi Shimbun* (Tokyo), April 3, 1990, opinion page. In Japanese.

———. "More Nuclear Power Means More Risk." *New York Times*, May 17, 2001, op-ed page. Available at http://www.nci.org/new/oped.htm.

———. "MOX Disposal of Surplus Weapons Plutonium: Politically Expedient, But Does It Make Sense?" Paper presented at the Fourth International Policy Forum: Management and Disposition of Nuclear Weapons Materials, Lansdowne, Va., February 12, 1997. Available at http://www.nci.org/s/sp21297.htm.

———. "The New Nuclear Threat." *Wall Street Journal*, June 8, 1994, op-ed page.

———. "The Nonproliferation Hoax." *New York Times*, August 20, 1990, op-ed page.

———. "Nuclear Deterrence and Nuclear Terrorism: Implications for Israel." Paper presented to the Israel Atomic Energy Commission and to the Tel Aviv University Conference on Terrorism, August 1988.

———. "The Nuclear Power and Nuclear Weapons Connection." *Social Education* (March 1990): 146–50.

———. "Nuclear Proliferation Overview: An Integrative Approach to Stopping the Spread of Nuclear Weapons." Paper presented to the W. Alton Jones Foundation, Charlottesville, Va., February 8, 1991.

———. "The Nuclear Watchdogs Have Failed." *International Herald Tribune*, September 24, 1991, op-ed page.

———. "Past, Present and Future." *Nuclear Engineering International* (September 2001): 39–40.

———. "Plugging the Leaks in Nuclear Export Controls: Why Bother?" *Orbis* (spring 1992): 167–80.

———. "Plutonium and the NPT." Paper presented at the Conference on Nuclear Non-Proliferation, Carnegie Endowment for International Peace, Washington, D.C., November 18, 1993.

———. "The Plutonium Industry and the Consequences for a Comprehensive Fissile Materials Cutoff." Paper presented at the Oxford Research Group Seminar, "Working Towards a Nuclear-Free World," in cooperation with Chinese People's Association for Peace & Disarmament, Oxford University, Oxford, England, April 28, 1997. Available at http://www.nci.org/s/sp42897.htm.

———. "A Pox on MOX." *Bulletin of the Atomic Scientists* 54 (2)(March–April 1998): 46–47. Available at http://www.nci.org/p/pl-bas98.htm.

———. "Reaching for a Common Ground." Paper presented to the International Policy Forum on Management & Disposition of Nuclear Weapons Material, Bethesda, Md., March 24, 1998. Available at http://www.nci.org/s/sp32498.htm.

———. "Ruminations on the Indian and Pakistani Nuclear Tests." *The Monitor* (University of Georgia Center for International Trade & Security) 4 (2–3)(spring–summer 1998): 3–5. Available at http://www.uga.edu/cits/publications/monitor_sp_su_1998.pdf.

———. "The Spread of Nuclear Weapons in the 1990s." *Medicine and War* 8 (October–December 1992): 259–62.

———. "Stabilization and Immobilization of Military Plutonium: A Non-Proliferation Perspective." Paper presented at the Plutonium Stabilization and Immobilization Workshop, U.S. Department of Energy, Washington, D.C., December 12, 1995. Available at http://www.nci.org/s/sp121295.htm.

———. "Upgrading Protection of Licensed Nuclear Power Reactors against Radiological Sabotage." Paper presented at the American Nuclear Society Working Conference on Nuclear Power Plant Security, Gaithersburg, Md., March 12, 1991.

———. "U.S.–Japan Accord Invites Proliferation." *Bulletin of the Atomic Scientists* (May 1988): 11–13.

———. "We Are Ignoring the Plutonium Issue at Our Peril." *Christian Science Monitor*, August 28, 1985, and *International Herald Tribune*, August 30, 1985, op-ed page.

———. "Weapons-Usable Nuclear Materials: Eliminate Them?" pp. 33–38. In *Director's Series on Proliferation*, edited by Kathleen Bailey. UCRL-LR-114070-1. Livermore, Calif.: Lawrence Livermore National Laboratory, June 7, 1993.

———. "What Should Be the Fundamental Basis of a National Plutonium Policy?" Paper presented at the International Policy Forum on Management and Disposition of Nuclear Weapons Materials, Leesburg, Va., March 8, 1994.

———. "What That Nuclear Pact with China Should Say." *Washington Post*, May 2, 1984, op-ed page.

Leventhal, Paul L., and Yonah Alexander, eds. *Nuclear Terrorism: Defining the Threat.* A Nuclear Control Institute book. Papers presented at the Conference on International Terrorism: The Nuclear Dimension, June 1985. Washington: Pergamon-Brassey's, 1986.

Leventhal, Paul L., and Yonah Alexander, eds. *Preventing Nuclear Terrorism: The Report and Papers of the International Task Force on Prevention of Nuclear Terrorism.* A Nuclear Control Institute book. Lexington, Mass.: Lexington Books, 1987.

Leventhal, Paul L., and Brahma Chellaney. "Nuclear Terrorism: Threat, Perception and Response in South Asia." Paper presented to the Institute for Defense Studies and Analyses, New Delhi, India, October 10, 1988. Reprinted in *Terrorism* 11 (1988). Available at http://www.nci.org/p/pl-bc.htm.

Leventhal, Paul L., and Steven Dolley. "Iraq Inspector Games." *Washington Post*, November 29, 1998, pp. C1 & C4. Available at http://www.nci.org/v-w-x/wp-jpg-1.htm.

———. "A Japanese Strategic Uranium Reserve: A Safe and Economic Alternative to Plutonium." Nuclear Control Institute, Washington, D.C., April 1993. Reprinted in *Science & Global Security* 5 (1994): 1–31.

———. "The MOX and Vitrification Options Compared: A Non-Proliferation Perspective." pp. 535–39. In *Proceedings of the Fifth International Conference on Radioactive Waste Management and Environmental Remediation: ICEM '95* (paper presented at the Fifth International Conference on Radioactive Waste Management and Environmental Remediation [ICEM '95], Berlin, Germany, September 3–7, 1995). Vol. 1. New York: American Society of Mechanical Engineers, 1995. Available at http://www.nci.org/b/berlin.htm.

———. "The North Korean Nuclear Crisis." *Medicine & Global Survival* 1(3)(September 1994): 164–75. Available at http://www.ippnw.org/MGS/V1N3Leventhal.html.

———. "The Plutonium Fallacy: An Update." Paper presented to the Special

Panel Session on Spent Fuel Reprocessing, Waste Management 99 Conference, Tucson, Arizona, March 1, 1999. Available at http://www.nci.org/p/plwm99.htm. Reprinted in *The Nonproliferation Review* 6 (3)(spring–summer 1999): 75–88. Available at http://cns.miis.edu/pubs/npr/vol06/63/leven63.pdf.

———. "Plutonium: The Heart of the Proliferation Threat in East Asia," pp. 80–104. In *Dealing with the North Korean Nuclear Problem*, edited by Taewoo Kim and Selig Harrison. Research Institute of Peace Studies, Peace Study Series No. 1. Seoul: Hanul Academy, 1995.

———. "Understanding Japan's Nuclear Transports: The Plutonium Context." Paper presented at the Conference on Ultrahazardous Radioactive Cargo by Sea: Implications and Responses, sponsored by the Maritime Institute of Malaysia, Kuala Lumpur, Malaysia, October 18, 1999. Available at http://www.nci.org/k-m/mmi.htm.

Leventhal, Paul L., and Eldon V. C. Greenberg. "The Renewal of Nuclear Trade with China: Legal and Policy Considerations," Nuclear Control Institute, Washington, D.C., November 21, 1996. Available at http://www.nci.org/i/ib112196.htm.

Leventhal, Paul L., Eldon V. C. Greenberg, Sharon Tanzer, and Steven Dolley. "Setting the Record Straight About Renegotiating the U.S.–EURATOM Nuclear Cooperation Agreement." Nuclear Control Institute, Washington, D.C., November 23, 1994. Available at http://www.nci.org/e/euribl.htm.

Leventhal, Paul L., and Milton Hoenig. "Mutually Assured Arms Reductions." *New York Times*, August 28, 1989, op-ed page.

———. "The Hidden Danger: Risks of Nuclear Terrorism." *Terrorism* 10 (1987): 1–22.

———. "Nuclear Installations and Potential Risks. The Hidden Danger: Risks of Nuclear Terrorism." Paper presented at the Parliamentary Assembly, Council of Europe, Paris, January 1987. Reprinted in *Terrorism* 10 (1987).

———. "Nuclear No Man's Land: Low Level Radioactive Wastes as an Unpoliced Diversion Path for Thefts of Weapons-Usable Nuclear Materials." Paper presented to the 2nd Committee of Investigation, Bundestag of the Federal Republic of Germany, Bonn. September 1988.

———. "Nuclear Terrorism: Reactor Sabotage and Weapons Proliferation Risks." Paper presented at the annual meeting of the Western Economic Association International, Lake Tahoe, Nev., June 22, 1989. Reprinted in *Contemporary Policy Issues* 8 (3)(July 1990): 106–21.

———. "The Tritium Factor." *New York Times*, August 4, 1987, op-ed page, and *International Herald Tribune*, August 5, 1987, op-ed page.

Leventhal, Paul L., Milton Hoenig, and Deborah Holland. "Crisis in the U.S. Nuclear Weapons Infrastructure," pp. 433–59. *Responding to a Changing World: Science and International Security*, edited by Eric Arnett. Washington, D.C.: American Association for the Advancement of Science, 1990.

———. "Japanese Plutonium Supply and Demand Requirements for Research,

Development and Demonstration Programs." Nuclear Control Institute, Washington, D.C., April 1990. Reprinted in the *Mainichi Economist*, April 10, 1990 (in Japanese).

Leventhal, Paul, Milton Hoenig, and Alan Kuperman. "Air Transport of Plutonium Obtained by the Japanese from Nuclear Fuel Controlled by the U.S." NCI Special Report, Nuclear Control Institute, Washington, D.C., March 1987.

Leventhal, Paul L., and Daniel Horner. "Peaceful Plutonium? No Such Thing." *New York Times*, January 25, 1995, op-ed page.

———. "The U.S.–China Nuclear Agreement: A Failure of Executive Policymaking and Congressional Oversight." *The Fletcher Forum* (winter 1987): 105–22.

Leventhal, Paul L., and Alan Kuperman. "Germany's Highly Enriched Uranium Decision at Garching: Impact on World Commerce in Bomb-Grade Uranium." Statement presented at Hearing of Fraktion Bundnis 90/Die Grunen on the Research Reactor Garching 2 (FRM-II), Bundestag, Bonn, Germany, November 1, 1995. Available at http://www.nci.org/s/sp11195.htm.

———. "RERTR at the Crossroads: Success or Demise?" Paper presented at the International RERTR Conference, Paris, September 18, 1995. Available at http://www.nci.org/s/sp91895.htm.

Leventhal, Paul L., and Edwin S. Lyman. "Who Says Iraq Isn't Making a Bomb?" *International Herald Tribune*, November 2, 1995, op-ed page. Available at http://www.nci.org/a/a11295.htm.

Leventhal, Paul L., and Sharon Tanzer, eds. *Averting a Latin American Nuclear Arms Race: New Prospects and Challenges for Argentine-Brazilian Nuclear Cooperation.* London: Macmillan, 1992.

———. "Fear and Folly in a Deadly Trade." *The Guardian*, October 4, 1991, op-ed page.

———. *The Tritium Factor: Tritium's Impact on Nuclear Arms Reductions.* Presentations and proceedings from the Tritium Factor Workshop, Cambridge, Mass., December 9, 1988, co-sponsored by Nuclear Control Institute and the American Academy of Arts and Sciences. Washington, D.C.: Nuclear Control Institute and the American Academy of Arts & Sciences, 1989.

Lyman, Edwin S. "Addressing Safety Issues in the Sea Transport of Radioactive Materials." Paper presented to Special Consultative Meeting, International Maritime Organization, London, March 4–6, 1996. Available at http://www.nci.org/i/ib31196a.htm.

———. "Comments." In "Report of the Expert Panel on the Role and Direction of Nuclear Regulatory Research." U.S. Nuclear Regulatory Commission, Washington, D.C., May 2001. Available at http://www.nci.org/e/el51001.htm.

———. "A Critique of Physical Protection Standards for Irradiated Materials." Paper presented at the 40th Annual Meeting of the Institute of Nuclear Materials Management (INMM), Phoenix, Ariz., July 1999. Available at http://www.nci.org/e/el-inmm99.htm.

———. "DOE Reprocessing Policy and the Irreversibility of Plutonium Disposition." *Proceedings of the 3rd Topical Meeting on DOE Spent Nuclear Fuel and Fissile Materials Management.* American Nuclear Society, Charleston, S.C., September 8–11, 1998. Available at http://www.nci.org/e/el-ans.htm.

———. "The Future of Immobilization Under the U.S.–Russian Plutonium Disposition Agreement." 42nd Annual Meeting of the Institute of Nuclear Materials Management (INMM), Indian Wells, Calif., July 18, 2001. Available at http://www.nci.org/new/el-inmm2001.htm.

———. "Inadequacy of the IAEA's Air Transport Regulations: The Case of MOX Fuel." Technical paper presented at the Dangerous Goods Panel, International Civil Aviation Organization (ICAO), Montreal, Canada, October 24, 1997.

———. "Japan's Plutonium Fuel Production Facility (PFPF): A Case Study of the Challenges of Nuclear Materials Management." Paper presented at the 39th Annual Meeting of the Institute of Nuclear Materials Management (INMM), Naples, Fla., July 1998.

———. "The Pebble-Bed Modular Reactor: Safety Issues." *Physics and Society* (October 2001). Available at http://www.aps.org/units/fps/oct01/a6oct01.pdf.

———. "Public Health Risks of Substituting Mixed-Oxide for Uranium Fuel in Pressurized Water Reactors." *Science and Global Security* 9 (2001): 1. Available at http://www.nci.org/l/lyman-mox-sgs.pdf.

———. "Safety Issues in the Sea Shipment of Vitrified High-Level Radioactive Wastes to Japan." Report sponsored by the Nuclear Control Institute, Greenpeace International, and Citizens' Nuclear Information Center, Tokyo, Japan, December 1994.

———. "The Sea Shipment of Radioactive Materials: Safety and Environmental Concerns." Paper presented at the Conference on Ultrahazardous Radioactive Cargo by Sea: Implications and Responses, sponsored by the Maritime Institute of Malaysia, Kuala Lumpur, Malaysia, October 1999. Available at http://www.nci.org/e/el-malaysia.htm.

———. "The Sea Transport of Vitrified High-Level Wastes: Unresolved Safety Issues." Report submitted to the 40th Session of the Marine Environment Protection Committee, International Maritime Organization, Washington, D.C., July 11, 1997.

———. "Weapons Plutonium: Just Can It." *Bulletin of the Atomic Scientists* 52 (6)(November–December 1996): 48–52.

Lyman, Edwin S., and Steven Dolley. "Accident Prone: The Trouble at Tokai-Mura." *Bulletin of the Atomic Scientists* (March/April 2000). Available at http://www.thebulletin.org/issues/2000/ma00/ma00lyman-dolley.html.

Lyman, Edwin, and Paul Leventhal. "Radiological Sabotage at Nuclear Power Plants: A Moving Target Set." Paper presented at the 41st Annual Meeting of the Institute of Nuclear Materials Management (INMM), New Orleans, La., July 2000. Available at http://www.nci.org/e/el-inmm2000.htm.

———. "Bury the Stuff." *Bulletin of the Atomic Scientists* 53 (2)(March–April 1997): 45–48. Available at http://www.nci.org/b/bas97.htm.

Mark, J. Carson. "Reactor-Grade Plutonium's Explosive Properties." Nuclear Control Institute, Washington, D.C., August 1990. Reprinted in *Science & Global Security* 4 (1993): 111–28.

Mark, J. Carson, Thomas D. Davies, Milton M. Hoenig, and Paul L. Leventhal. "The Tritium Factor as a Forcing Function in Nuclear Arms Reduction Talks." *Science* 241 (September 2, 1988): 1166–69.

Mark, J. Carson, Theodore Taylor, Eugene Eyster, William Maraman, and Jacob Wechsler. "Can Terrorists Build Nuclear Weapons?" pp. 55–65. In *Preventing Nuclear Terrorism*, edited by Paul Leventhal and Yonah Alexander. Lexington, Mass.: Lexington Books, 1987.

Miller, Marvin M. "Are IAEA Safeguards on Plutonium Bulk-Handling Facilities Effective?" Nuclear Control Institute, Washington, D.C., 1990.

Patterson, Walter C. *The Plutonium Business*. A Nuclear Control Institute book. San Francisco: Sierra Club Books and London: Paladin Books, 1984.

Redick, John R. "Argentina and Brazil's New Arrangement for Mutual Inspections and IAEA Safeguards." Nuclear Control Institute, Washington, D.C., February 1992.

Van Dyke, Jon M. "Applying the Precautionary Principle to Ocean Shipments of Radioactive Materials." Paper presented at the Special Consultative Meeting, International Maritime Organization, London, March 4–6, 1996. Available at http://www.nci.org/i/ib3496a.htm.

———. "The Legitimacy of Unilateral Actions to Protest the Ocean Shipment of Ultrahazardous Radioactive Materials." Nuclear Control Institute, Washington, D.C., December 13, 1996. Available at http://www.nci.org/i/ib121396.htm.

———. "The Need for Further International Action Regarding Safety of Sea Transport of Ultrahazardous Radioactive Materials." Nuclear Control Institute, Washington, D.C., August 17, 1998. Available at http://www.nci.org/v-w-x/vd-81798.htm.

Van Dyke, Jon M., and Eldon V. C. Greenberg. "International Law Permits Panama to Prohibit Shipments of Ultrahazardous Radioactive Materials through the Panama Canal." Nuclear Control Institute, Washington, D.C., January 12, 2000. Available at http://www.nci.org/v-w-x/vd-eg-canal.htm.

Working Group on Nuclear Explosives Control Policy. "Stopping the Spread of Nuclear Weapons: Assessment of Current Policy. Agenda for Action." Nuclear Control Institute, Washington, D.C., July 1984.

ABOUT THE CONTRIBUTORS AND EDITORS

ZACHARY S. DAVIS

Zachary S. Davis is an analyst of foreign nuclear programs at the Z Division of Lawrence Livermore National Laboratory, with special knowledge of the nuclear programs of India, Pakistan, and North Korea. He was a nonproliferation policy analyst for the Congressional Research Service for ten years, where he worked with key congressional committees to develop nonproliferation, arms control, export control, and sanctions legislation. He also served in the U.S. Department of State, Office of the Deputy Secretary, implementing the U.S. response to India and Pakistan's nuclear tests. At the Arms Control and Disarmament Agency, he worked on strengthening safeguards to improve International Atomic Energy Agency inspections. He is the author of numerous government reports on foreign nuclear programs and government studies. He received M.A. and Ph.D. degrees in international relations from the University of Virginia and a B.A. in politics from the University of California–Santa Cruz.

STEVEN DOLLEY

Steven Dolley is the research director of the Nuclear Control Institute, having joined the institute staff in June 1991. Mr. Dolley monitors civilian plutonium programs in Europe, Japan, and China; nuclear weapon development programs in Iraq and North Korea; and U.S. government programs for disposing of excess military plutonium. He is the author of numerous articles in such publications as the *Bulletin of Atomic Scientists, Science and Global Security, Nuclear Engineering International,* and *Journal of the American Forensic Association*, and he prepares the institute's background papers and issue briefs. From 1988 to 1991 Mr. Dolley was an instructor in communications and debate at Bates College and the University of Vermont, where his principal focus was arms control. He was a 1984 honors graduate of Bates College and pursued a master's degree in communications at the University of North Carolina at Chapel Hill.

HAROLD A. FEIVESON

Harold A. Feiveson is senior research policy scientist at Princeton University's Center for Energy and Environmental Studies. Dr. Feiveson's principal research interests are in the fields of nuclear weapons and nuclear energy policy. He is a coprincipal investigator of Princeton's Research Program on Nuclear Policy Alternatives. His recent work has focused on the ways in which the nuclear arsenals of the United States and the former Soviet Union can be dismantled and "de-alerted," on the strengthening of the nuclear nonproliferation regime (including a universal ban on the production of weapon-usable material and on nuclear weapon testing), and on the strengthening of the separation between nuclear weapons and civilian nuclear energy activities. He is the editor of *The Nuclear Turning Point: A Blueprint for Deep Cuts and De-Alerting of Nuclear Weapons* and author of the "Overview" and "Verification" chapters (Washington, D.C.: Brookings Institution, 1999). Dr. Feiveson has an M.S. degree in theoretical physics from the University of California–Los Angeles, 1959, and a Ph.D. in public affairs from Princeton University, 1972.

ROBERT L. GALLUCCI

Robert L. Gallucci was appointed dean of Georgetown University's Edmund A. Walsh School of Foreign Service in 1996. He had just completed twenty-one years of government service, serving since August 1994 with the U.S. Department of State as ambassador at large. In 1998 he became the State Department's special envoy to deal with the threat posed by the proliferation of ballistic missiles and weapons of mass destruction. In 1991 he was appointed deputy executive chairman of the UN Special Commission overseeing the disarmament of Iraq. In 1992 he became senior coordinator for nonproliferation and nuclear safety initiatives in the former Soviet Union. Later in 1992 Dr. Gallucci was confirmed as the assistant secretary of state for political-military affairs. He has authored a number of publications on political-military issues, including *Neither Peace nor Honor: The Politics of American Military Policy in Vietnam.* He holds a bachelor's degree from the State University of New York at Stony Brook, and a master's and doctorate in politics from Brandeis University. Before joining the State Department, he taught at Swarthmore College, Johns Hopkins School for Advanced International Studies, and Georgetown University.

RICHARD L. GARWIN

Richard L. Garwin is Philip D. Reed senior fellow for science and technology at the Council on Foreign Relations, New York, and IBM fellow emeritus at the Thomas J. Watson Research Center, Yorktown Heights, New York. He joined IBM Corporation in 1952, after three years on the faculty of the University of Chicago, and was until June 1993 IBM fellow at the Thomas J. Watson Research Center. He has also been professor of public policy in the Kennedy School of Government, Harvard University. Dr. Garwin received his Ph.D. in physics from the University of Chicago in 1949. Dr. Garwin is coauthor of many books, among them *Nuclear Weapons and World Politics*, *Nuclear Power Issues and*

Choices, Energy: The Next Twenty Years, Science Advice to the President, and *Managing the Plutonium Surplus: Applications and Technical Options.* He is a fellow of the American Physical Society and of the American Academy of Arts and Sciences and a member of the National Academy of Sciences. He is a longtime member of Pugwash and has served on the Pugwash Council. The U.S. government awarded him the 1996 R. V. Jones Foreign Intelligence Award and the 1996 Enrico Fermi Award.

JAMES A. HASSBERGER

James A. Hassberger has been with the Lawrence Livermore National Laboratory for more than sixteen years. He is currently a senior nuclear scientist with the Energy and Environmental Directorate, where he supports efforts by the laboratory and the U.S. Department of Energy in nuclear energy and security issues, including nonproliferation and interactions with the former Soviet Union. Prior to moving to Livermore, Mr. Hassberger was a senior nuclear engineer with Westinghouse, working in both fast reactor and fusion energy programs. He received his master's degree in nuclear engineering from the University of Michigan in 1972.

WILLIAM LANOUETTE

William Lanouette, a writer and public policy analyst, has studied and written about nuclear weapons and nuclear power for more than thirty years. He is the author of *Genius in the Shadows: A Biography of Leo Szilard, The Man Behind the Bomb.* He holds an A.B. in English from Fordham College and an M.Sc. and Ph.D. in politics from the London School of Economics (University of London). A staff member for *Newsweek* and the *National Journal,* and Washington correspondent for *The Bulletin of the Atomic Scientists,* he has also written articles for *Arms Control Today, Atlantic Monthly, Economist, Energy Daily, Scientific American, Smithsonian, Washington Post,* and *Wilson Quarterly.* Since 1991 he has been a senior policy analyst for energy and science issues at the U.S. General Accounting Office in Washington, D.C.

PAUL L. LEVENTHAL

Paul L. Leventhal is president of the Nuclear Control Institute, which he founded in 1981 after holding senior staff positions in the U.S. Senate. He has prepared four books for the Nuclear Control Institute and lectured in a number of countries on nuclear issues. He came to Washington, D.C., in 1969 as press secretary to Senator Jacob K. Javits (R-N.Y.) after a decade of political and investigative reporting for the *Cleveland Plain Dealer, New York Post,* and *Newsday.* In 1972 he served as congressional correspondent for the *National Journal.* He returned to Capitol Hill as special counsel to the Senate Government Operations Committee, 1972–1976, and also served as staff director of the Senate Nuclear

Regulation Subcommittee, 1979–1981. He contributed to enactment of the Energy Reorganization Act of 1974 and the Nuclear Non-Proliferation Act of 1978. He was codirector of the Senate Special Investigation of the Three Mile Island nuclear accident, 1979–1980, and helped prepare the "lessons-learned" legislation enacted in 1980. He was research fellow at Harvard University's Program for Science and International Affairs, 1976–1977. He served as assistant administrator for policy and planning at the U.S. National Oceanic and Atmospheric Administration, 1977–1978. He holds a bachelor's degree in government, magna cum laude, from Franklin and Marshall College, which presented him its Alumni Medal in 1988, and a master's degree from the Columbia University Graduate School of Journalism.

AMORY B. LOVINS

Amory B. Lovins cofounded Rocky Mountain Institute in 1982 and serves as its chief executive officer (research). An experimental physicist educated at Harvard and Oxford, Mr. Lovins rose to prominence during the oil crises of the 1970s when, still in his twenties, he challenged conventional supply-side dogma by urging that the United States instead follow a "soft energy path." His controversial recommendations were eventually accepted by the energy industry, and his book, *Soft Energy Paths: Toward a Durable Peace*, went on to inspire a generation of decisionmakers. Lovins has received six honorary doctorates, a MacArthur Fellowship, and the Heinz, Lindbergh, and Right Livelihood awards, among other honors. His work today focuses on transforming the car, real estate, electricity, water, semiconductor, and several other manufacturing sectors toward advanced resource productivity. He has briefed twelve heads of state, held several visiting academic chairs, authored or coauthored twenty-seven books, and consulted for scores of industries and governments worldwide. The *Wall Street Journal* named Mr. Lovins one of thirty-nine people worldwide "most likely to change the course of business in the '90s."

EDWIN S. LYMAN

Edwin S. Lyman became scientific director of the Nuclear Control Institute in 1995. His research focuses on security and environmental issues associated with the management of nuclear materials. He has published numerous articles in journals, including *The Bulletin of the Atomic Scientists* and *Science and Global Security*. He is active as a member of the Institute of Nuclear Materials Management. From 1997 to 1998 he participated in the Processing Needs Assessment conducted by the U.S. Department of Energy's Nuclear Material Stabilization Task Group. He received a doctorate in theoretical physics from Cornell University in 1992, where he was an A. D. White Scholar. He was a postdoctoral research associate at Princeton University's Center for Energy and Environmental Studies from 1992 to 1995.

REPRESENTATIVE EDWARD J. MARKEY

Edward J. Markey, a Democrat, was elected to the U.S. House of Representatives in 1976 from a district that includes both blue-collar and high-tech suburbs north and west of downtown Boston. He serves on the House Budget and Commerce Committees. He is cochair with Representative Chris Shays (R-CT) of the Bipartisan Task Force on Non-Proliferation to help educate their congressional colleagues on the spread of nuclear weapons and other weapons of mass destruction and on the urgent need to control and reverse that trend. He is a veteran of congressional battles for the nuclear freeze, the Nuclear Test Ban Treaty, restrictions on resale of nuclear technology by China and North Korea, and maintenance of a wall between nuclear weapons and civilian nuclear power at home and abroad. He is now leading the effort in Congress for mutual U.S.–Russian cooperation to de-alert their nuclear weapons. He is also a leader on telecommunications policy, consumer rights, health reform, elimination of large monopolies, and conservation of energy and environmental resources. He has worked closely with the Nuclear Control Institute, Conservation Law Foundation, Consumer Federation of America, Children's Defense Fund, and National Education Association. He attended Boston College (B.A., 1968) and Boston College Law School (J.D., 1972). He served in the U.S. Army Reserve and was elected to the Massachusetts State House, where he served two terms.

MARVIN MILLER

Marvin Miller retired from the position of senior research scientist in the Massachusetts Institute of Technology Department of Nuclear Engineering in 1996. He is now senior scientist emeritus at both the institute's Center for International Studies, where he is a member of the Security Studies Program, and the Nuclear Engineering Department. Trained as a physicist, he was a tenured associate professor of electrical engineering at Purdue University, conducting research on laser theory and applications before joining Massachusetts Institute of Technology in 1976. At the institute his principal activities have been in the areas of nuclear arms control and the environmental impacts of energy use. In arms control, his major regional interests are the Middle East and South Asia. From 1984 to 1986 Dr. Miller was on leave from the Massachusetts Institute of Technology to serve with the U.S. Arms Control and Disarmament Agency, and, after returning to the institute, he continues in that capacity with the State Department. He also has been a consultant for the International Atomic Energy Agency, U.S. Department of Energy, and Los Alamos and Lawrence Livermore National Laboratories. He is the author of more than eighty publications in the fields of laser theory and applications, energy, and arms control.

GEORGE PERKOVICH

George Perkovich is the author of *India's Nuclear Bomb*, which received the American Historical Association's Herbert Feis Award for outstanding work by an independent scholar. His writing has appeared in *Foreign Affairs* and other publications. He has testified before both houses of Congress on South Asian security affairs and served on the 1997 Council on Foreign Relations Task Force that published "A New U.S. Policy toward India and Pakistan." He was deputy director for programs and director of the Secure World Program of the W. Alton Jones Foundation, a $400 million philanthropic institution located in Charlottesville, Va. In addition to managing the Secure World Program's $11 million annual grantmaking budget and designing and implementing initiatives to further the foundation board's mandate of reducing the risk of nuclear war, he oversaw the $14 million Sustainable World Program. Perkovich received his B.A. in politics from the University of California–Santa Cruz in 1980, his M.A. in Soviet studies from Harvard University in 1986, and his Ph.D. in foreign affairs from the University of Virginia in 1997.

RICHARD RHODES

Richard Rhodes is the author of eighteen books, including *The Making of the Atomic Bomb*, which won a Pulitzer Prize in nonfiction, a National Book Award, and a National Book Critics Circle Award; *Dark Sun: The Making of the Hydrogen Bomb*, which was short-listed for a Pulitzer Prize in history; *Nuclear Renewal: Common Sense About Energy*; and, most recently, *Why They Kill: The Discoveries of a Maverick Criminologist.* He has received numerous fellowships for research and writing, including grants from the Ford Foundation, Guggenheim Foundation, MacArthur Foundation, and Alfred P. Sloan Foundation. He has been a visiting scholar at Harvard University and the Massachusetts Institute of Technology and a host and correspondent for documentaries on nuclear issues on public television's *Frontline* and *American Experience* series. A Kansas native and a 1959 honors graduate of Yale University, he lives in rural Connecticut.

LAWRENCE SCHEINMAN

Lawrence Scheinman is distinguished professor of international policy at the Monterey Institute of International Studies, emeritus professor at Cornell University, and adjunct professor at Georgetown University. He also has been a member of the tenured faculties at the University of Michigan and the University of California–Los Angeles. His government service includes appointment as assistant director of the U.S. Arms Control and Disarmament Agency, responsible for nonproliferation and regional arms control during the administration of President William J. Clinton, and earlier appointments in the Department of Energy, Department of State, and Energy Research and Development Administration. He also served for two years as special assistant to Director General Hans Blix at the International Atomic Energy Agency. Dr. Scheinman has published extensively on nuclear proliferation, arms control, safeguards, international relations, and international organization. He is a member of the Council on Foreign Relations and of the State Department Arms Control and Non-Proliferation Advisory Board. He is admitted to practice before the Bar of the State of New York.

SHARON TANZER

Sharon Tanzer is vice president of the Nuclear Control Institute and a member of its board of directors. She was co-editor of *Averting a Nuclear Arms Race in Latin America* (Macmillan, 1991), the proceedings of the institute's 1989 Montevideo conference. She was rapporteur and editor of *The Tritium Factor* (Nuclear Control Institute and American Academy of Arts and Sciences, 1989), the proceedings of a 1988 nuclear arms control workshop cosponsored by the Nuclear Control Institute and the academy. She was also project coordinator for the institute's International Task Force on Prevention of Nuclear Terrorism. She has organized and participated in the institute's meetings and briefings on nuclear proliferation and nuclear terrorism issues in Europe, India, Japan, and Latin America. Ms. Tanzer received a B.A. in history, cum laude, from Barnard College, was a Fulbright fellow at the University of Bordeaux, France, and earned a master's degree in history from Stanford University.

ROBERT H. WILLIAMS

Robert H. Williams is senior research scientist at Princeton University's Center for Energy and Environmental Studies. His research interests span a wide range of topics relating to advanced energy technologies, energy strategies, and energy policy for both industrialized and developing countries. A considerable part of his research is focused on energy technologies and strategies for developing countries, where most of the growth in global energy demand will take place, and where environmental and security challenges relating to energy are especially great. He was chair of the Renewable Energy Task Force for the President's Committee of Advisors on Science and Technology and was the principal author of chapter 6, "Renewable Energy," in the 1997 report, *Federal Energy Research & Development for the Challenges of the 21st Century and Report of the Energy R&D Panel.* He received a B.S. in physics from Yale University in 1962 and a Ph.D. in theoretical plasma physics from the University of California at Berkeley in 1967.

BERTRAM WOLFE

Dr. Bertram Wolfe retired from General Electric in 1992 as a vice president and general manager of the company's Nuclear Energy Business after a career of more than thirty-five years with General Electric. Dr. Wolfe has worked in nearly all technical phases of peaceful nuclear power and has had responsibility for a number of successful nuclear reactor projects. In 1987 he was appointed a vice president of General Electric and manager of its Nuclear Energy Division. He is presently an independent consultant in the fields of business, energy, and nuclear energy. Dr. Wolfe was elected to the National Academy of Engineering in 1980. He was elected president of the American Nuclear Society in 1986–87; was the recipient of the Walter Zinn Technical Accomplishment Award in 1990; was honored with the Henry DeWolf Smyth Nuclear Statesman Award in 1992; and was presented with the Tommy Thompson Nuclear Safety Award in 1997. He is a fellow of the American Nuclear Society. Dr. Wolfe received a B.A. in physics from Princeton University and a Ph.D. in nuclear physics from Cornell University.